Honda CB250RS Owners Workshop Manual

by Pete Shoemark
With an additional Chapter on the CB250 RSD-C model
by Jeremy Churchill

CB250 RS-A. 248cc. April 1980 to March 1983
CB250 RSD-C. 248cc. January 1982 to 1984

Note: This manual covers all SOHC engine models

ISBN 978 1 85010 144 4

(932-7Q5)

J H Haynes & Co. Ltd.
Haynes North America, Inc

www.haynes.com

British Library Cataloguing in Publication Data

Shoemark, Pete
Honda CB250 RS singles owners workshop manual.—
2nd ed.—(Haynes owners workshop manuals)
1. Honda motorcycle
I. Title II. Churchill, Jeremy III. Series
629.28'775 TL448.H6
ISBN 1–85010–144–2

Acknowledgements

Our thanks are due to Fran Ridewood and Co of Wells who supplied both machines featured in this manual and to Paul Branson Motorcycles of Yeovil, who supplied much of the service information.

The Avon Rubber Company provided advice and information on tyre fittings; NGK Spark Plugs (UK) Ltd furnished advice about sparking plug conditions; and Renold Ltd provided information relating to chain care and renewal.

About this manual

The purpose of this manual is to present the owner with a concise and graphic guide which will enable him to tackle any operation from basic routine maintenance to a major overhaul. It has been assumed that any work would be undertaken without the luxury of a well-equipped workshop and a range of manufacturer's service tools.

To this end, the machine featured in the manual was stripped and rebuilt in our own workshop, by a team comprising a mechanic, a photographer and the author. The resulting photographic sequence depicts events as they took place, the hands shown being those of the author and the mechanic.

The use of specialised, and expensive, service tools was avoided unless their use was considered to be essential due to risk of breakage or injury. There is usually some way of improvising a method of removing a stubborn component, provided that a suitable degree of care is exercised.

The author learnt his motorcycle mechanics over a number of years, faced with the same difficulties and using similar facilities to those encountered by most owners. It is hoped that this practical experience can be passed on through the pages of this manual.

Where possible, a well-used example of the machine is chosen for the workshop project, as this highlights any areas which might be particularly prone to giving rise to problems. In this way, any such difficulties are encountered and resolved before the text is written, and the techniques used to deal with them can be incorporated in the relevant sections. Armed with a working knowledge of the machine, the author undertakes a considerable amount of reseach in order that the maximum amount of data can be included in this manual.

Each Chapter is divided into numbered sections. Within these sections are numbered paragraphs. Cross reference throughout the manual is quite straightforward and logical. When reference is made 'See Section 6.10' it means Section 6, paragraph 10 in the same Chapter. If another Chapter were intended the reference would read, for example, 'See Chapter 2, Section 6.10'. All the photographs are captioned with a section/paragraph number to which they refer and are relevant to the Chapter text adjacent.

Figures (usually line illustrations) appear in a logical but numerical order, within a given Chapter. Fig. 1.1 therefore refers to the first figure in Chapter 1.

Left-hand and right-hand descriptions of the machines and their components refer to the left and right of a given machine when the rider is seated normally.

Motorcycle manufacturers continually make changes to specifications and recommendations, and these, when notified, are incorporated into our manuals at the earliest opportunity.

Contents

1981 Honda CB250 RS

Engine/gearbox unit - right-hand view

Engine/gearbox unit - left-hand view

Introduction to the Honda CB250 RS

The Honda CB250 RS is one of the lightest and smallest 250cc four-stroke motorcycles available today. It is powered by a four-valve overhead camshaft (ohc) engine which is derived from the unit fitted to the XL250 S trail bike. The engine incorporates a balancer mechanism which counteracts the inherent imbalance found in all single cylinder engines. This in turn permits the use of a lightweight frame which utilises the engine itself as a stressed member. To reduce weight further, non-essentials such as an electric starter have been omitted, the end result being a dry weight of 276 lb (125 kg); well below the norm for this capacity of Japanese motorcycle.

Front suspension is by conventional oil-damped telescopic forks, whilst the rear swinging arm unit is supported by FVQ gas/oil suspension units incorporating spring preload adjustment. The fuel tank is styled to blend into a continuous unit with the side panels, seat and tail hump, emphasising the lightweight theme. Extensive use of plastics in the mudguards, tail hump, side panels and the chain and sprocket guards make for less weight and cost.

Braking is by a single hydraulic disc at the front and a drum brake at the rear. Both wheels are of the wire spoked type using light alloy rims. The 12 volt electrical system is powered by a crankshaft-mounted alternator controlled by a sealed solid-state rectifier/regulator unit. Ignition is of the non-adjustable capacitor discharge (CDI) type.

The overall result is a light, low maintenance machine ideally suited for commuting, but with the added appeal of sporting inclinations certain to gain the enthusiasm of younger riders.

Model dimensions and weight

CB250 RS

Overall length .... 2055 mm (80.9 in)

Overall width .... 755 mm (29.7 in)

Overall height .... 1115 mm (43.9 in)

Wheelbase .... 1350 mm (53.2 in)

Dry weight .... 125 kg (276 lb)

Ordering Spare Parts

When ordering spare parts for any Honda model it is advisable to deal direct with an official Honda agent, who should be able to supply most items ex-stock. Parts cannot be obtained from Honda (UK) Limited direct; all orders must be routed via an approved agent, even if the parts required are not held in stock.

Always quote the engine and frame numbers in full. The frame number is located on the left-hand side of the steering head and the engine number is stamped on the upper crankcase, immediately to the rear of the cylinder. Use only parts of genuine Honda manufacture. Pattern parts are available, some of which originate from Japan, but in many instances, they may have an adverse effect on performance and/or reliability. Furthermore the fitting of non-standard parts may invalidate the warranty. Honda do not operate a 'service exchange' scheme.

Some of the more expendable parts such as sparking plugs, bulbs, tyres, oils and greases etc., can be obtained from accessory shops and motor factors, who have convenient opening hours, and can often be found not far from home. It is also possible to obtain parts on a Mail Order basis from a number of specialists who advertise regularly in the motor cycle magazines.

Frame number location

Engine number location

Safety first!

Professional motor mechanics are trained in safe working procedures. However enthusiastic you may be about getting on with the job in hand, do take the time to ensure that your safety is not put at risk. A moment's lack of attention can result in an accident, as can failure to observe certain elementary precautions.

There will always be new ways of having accidents, and the following points do not pretend to be a comprehensive list of all dangers; they are intended rather to make you aware of the risks and to encourage a safety-conscious approach to all work you carry out on your vehicle.

Essential DOs and DON'Ts

DON'T start the engine without first ascertaining that the transmission is in neutral.

DON'T suddenly remove the filler cap from a hot cooling system – cover it with a cloth and release the pressure gradually first, or you may get scalded by escaping coolant.

DON'T attempt to drain oil until you are sure it has cooled sufficiently to avoid scalding you.

DON'T grasp any part of the engine, exhaust or silencer without first ascertaining that it is sufficiently cool to avoid burning you.

DON'T allow brake fluid or antifreeze to contact the machine's paintwork or plastic components.

DON'T syphon toxic liquids such as fuel, brake fluid or antifreeze by mouth, or allow them to remain on your skin.

DON'T inhale dust – it may be injurious to health (see *Asbestos* heading).

DON'T allow any spilt oil or grease to remain on the floor – wipe it up straight away, before someone slips on it.

DON'T use ill-fitting spanners or other tools which may slip and cause injury.

DON'T attempt to lift a heavy component which may be beyond your capability – get assistance.

DON'T rush to finish a job, or take unverified short cuts.

DON'T allow children or animals in or around an unattended vehicle.

DON'T inflate a tyre to a pressure above the recommended maximum. Apart from overstressing the carcase and wheel rim, in extreme cases the tyre may blow off forcibly.

DO ensure that the machine is supported securely at all times. This is especially important when the machine is blocked up to aid wheel or fork removal.

DO take care when attempting to slacken a stubborn nut or bolt. It is generally better to pull on a spanner, rather than push, so that if slippage occurs you fall away from the machine rather than on to it.

DO wear eye protection when using power tools such as drill, sander, bench grinder etc.

DO use a barrier cream on your hands prior to undertaking dirty jobs – it will protect your skin from infection as well as making the dirt easier to remove afterwards; but make sure your hands aren't left slippery. Note that long-term contact with used engine oil can be a health hazard.

DO keep loose clothing (cuffs, tie etc) and long hair well out of the way of moving mechanical parts.

DO remove rings, wristwatch etc, before working on the vehicle – especially the electrical system.

DO keep your work area tidy – it is only too easy to fall over articles left lying around.

DO exercise caution when compressing springs for removal or installation. Ensure that the tension is applied and released in a controlled manner, using suitable tools which preclude the possibility of the spring escaping violently.

DO ensure that any lifting tackle used has a safe working load rating adequate for the job.

DO get someone to check periodically that all is well, when working alone on the vehicle.

DO carry out work in a logical sequence and check that everything is correctly assembled and tightened afterwards.

DO remember that your vehicle's safety affects that of yourself and others. If in doubt on any point, get specialist advice.

IF, in spite of following these precautions, you are unfortunate enough to injure yourself, seek medical attention as soon as possible.

Asbestos

Certain friction, insulating, sealing, and other products – such as brake linings, clutch linings, gaskets, etc – contain asbestos. *Extreme care must be taken to avoid inhalation of dust from such products since it is hazardous to health.* If in doubt, assume that they *do* contain asbestos.

Fire

Remember at all times that petrol (gasoline) is highly flammable. Never smoke, or have any kind of naked flame around, when working on the vehicle. But the risk does not end there – a spark caused by an electrical short-circuit, by two metal surfaces contacting each other, by careless use of tools, or even by static electricity built up in your body under certain conditions, can ignite petrol vapour, which in a confined space is highly explosive.

Always disconnect the battery earth (ground) terminal before working on any part of the fuel or electrical system, and never risk spilling fuel on to a hot engine or exhaust.

It is recommended that a fire extinguisher of a type suitable for fuel and electrical fires is kept handy in the garage or workplace at all times. Never try to extinguish a fuel or electrical fire with water.

Note: *Any reference to a 'torch' appearing in this manual should always be taken to mean a hand-held battery-operated electric lamp or flashlight. It does* ***not*** *mean a welding/gas torch or blowlamp.*

Fumes

Certain fumes are highly toxic and can quickly cause unconsciousness and even death if inhaled to any extent. Petrol (gasoline) vapour comes into this category, as do the vapours from certain solvents such as trichloroethylene. Any draining or pouring of such volatile fluids should be done in a well ventilated area.

When using cleaning fluids and solvents, read the instructions carefully. Never use materials from unmarked containers – they may give off poisonous vapours.

Never run the engine of a motor vehicle in an enclosed space such as a garage. Exhaust fumes contain carbon monoxide which is extremely poisonous; if you need to run the engine, always do so in the open air or at least have the rear of the vehicle outside the workplace.

The battery

Never cause a spark, or allow a naked light, near the vehicle's battery. It will normally be giving off a certain amount of hydrogen gas, which is highly explosive.

Always disconnect the battery earth (ground) terminal before working on the fuel or electrical systems.

If possible, loosen the filler plugs or cover when charging the battery from an external source. Do not charge at an excessive rate or the battery may burst.

Take care when topping up and when carrying the battery. The acid electrolyte, even when diluted, is very corrosive and should not be allowed to contact the eyes or skin.

If you ever need to prepare electrolyte yourself, always add the acid slowly to the water, and never the other way round. Protect against splashes by wearing rubber gloves and goggles.

Mains electricity and electrical equipment

When using an electric power tool, inspection light etc, always ensure that the appliance is correctly connected to its plug and that, where necessary, it is properly earthed (grounded). Do not use such appliances in damp conditions and, again, beware of creating a spark or applying excessive heat in the vicinity of fuel or fuel vapour. Also ensure that the appliances meet the relevant national safety standards.

Ignition HT voltage

A severe electric shock can result from touching certain parts of the ignition system, such as the HT leads, when the engine is running or being cranked, particularly if components are damp or the insulation is defective. Where an electronic ignition system is fitted, the HT voltage is much higher and could prove fatal.

Routine maintenance

Periodical routine maintenance is essential to keep the motorcycle in a safe condition and at peak performance. Routine maintenance also saves money because it provides the opportunity to detect and remedy a fault before it develops further and causes more damage. Maintenance should be undertaken on either a calendar or mileage basis depending on whichever comes sooner. The period between maintenance tasks serves only as a guide since there are many variables eg; age of machine, riding technique and adverse conditions.

The maintenance instructions are generally those recommended by the manufacturer but are supplemented by additional tasks which, through practical experience, the author recommends should be caried out at the intervals suggested. The additional tasks are primarily of a preventative nature, which will assist in eliminating unexpected failure of a component or system, due to wear and tear, and increase safety margins when riding.

All the maintenance tasks are described in detail together with the procedures required for accomplishing them. If necessary, more general information on each topic can be found in the relevant Chapter within the main text.

Although no special tools are required for routine maintenance, a good selection of general workshop tools is essential. Included in the tools must be a range of metric ring or combination spanners, a selection of crosshead screwdrivers, and two pairs of circlip pliers, one external opening and the other internal opening. Additionally, owing to the extreme tightness of most cross-head screws on Japanese machines, an impact screwdriver, together with a choice of large or small cross-head screw bits, is absolutely indispensable. This is particularly so if the engine has not been dismantled since leaving the factory.

Weekly, or every 200 miles (320 km)

1 Tyres

Check the tyre pressures. Always check the pressure when the tyres are cold as the heat generated when the machine has been ridden can increase the pressure by as much as 8 psi, giving a totally inaccurate reading. Variations in pressure of as little as 2 psi may alter certain handling characteristics. It is therefore recommended that whatever type of pressure gauge is used, it should be checked occasionally to ensure accurate readings. Do not put absolute faith in 'free air' gauges at garages or petrol stations. They have been known to be in error.

Inspect the tyre treads for cracking or evidence that the outer rubber is leaving the inner cover. Also check the tyre walls for splitting or perishing. Carefully inspect the treads for stones, flints or shrapnel which may have become embedded and be slowly working their way towards the inner surface. Remove such objects with a suitable tool.

Tyre pressures (cold)

	Front	*Rear*
Solo	*24 psi (1.75 kg/cm^2)*	*32 psi (2.25 kg/cm^2)*
With passenger	*24 psi (1.75 kg/cm^2)*	*36 psi (2.50 kg/cm^2)*

2 Checking the engine/gearbox oil level

Place the machine securely on its centre stand on level ground. Unscrew the oil filler plug from the front of the right-hand engine cover, and wipe the integral dipstick on a piece of clean rag. Place the dipstick/plug unit back in the filler hole, but do not screw it home. The oil level measurement is made with the dipstick just resting against the filler hole. Remove the dipstick and note the oil level reading. The upper and lower limits are marked by horizontal ribs with a diagonal hatched area between them. If necessary, top up the oil using a good quality SAE 10W/40 engine oil. Note that the oil level should not be checked immediately after riding, because a proportion of the oil will cling to the internal surfaces of the engine and will require a few minutes to drain down into the sump.

3 Brake fluid

Check the hydraulic fluid level in the front brake master cylinder reservoir. Before removing the reservoir cap and diaphragm place the handlebars in such a position that the reservoir is approximately vertical. This will prevent spillage. The fluid should lie between the upper and lower lines on the reservoir body. Replenish, if necessary, with hydraulic brake fluid of the correct specification, which is DOT 3 (USA) or SAE-J1703. If the level of fluid in the reservoir is excessively low, check the pads for wear. If the pads are not worn, suspect a fluid leakage in the system. This must be rectified immediately. Refer to Chapter 5 for further information.

Monthly or every 600 miles (1000 km)

Complete the operations listed under the weekly/200 mile heading, then carry out the following:

1 Battery electrolyte level

Access to the Yuasa battery is gained after removing the right-hand side panel. The electrolyte level can be checked visually through the battery's transparent case. Make sure that the level in each cell is between the minimum and maximum lines on the battery case and that the vent pipe has not become pinched or obstructed. The transparent case also makes it possible for a quick check on the condition of the battery plates and separators.

Unless acid is spilt, as may occur if the machine falls over, the electrolyte should always be topped up with distilled water to restore the correct level. If acid is spilt on any part of the machine, it should be neutralised with an alkali such as washing soda, and washed away with plenty of water, otherwise serious corrosion will occur. Top up with sulphuric acid, only in the event of spillage.

2 *Lubricating the controls, cables and pivots*

Clean and examine the various control lever and pedal pivots, lubricating each one with light machine oil or one of the multi-purpose maintenance aerosols. Check the outer cables for signs of damage, then examine the exposed portions of the inner cables. Any signs of kinking or fraying will indicate that renewal is required. To obtain maximum life and reliability from the cables they should be thoroughly lubricated. To do the job properly and quickly use one of the hydraulic cable oilers available from most motorcycle shops. Free one end of the cable and assemble the cable oiler as described by the manufacturer's instructions. Operate the oiler until oil emerges from the lower end, indicating that the cable is lubricated throughout its length. This process will expel any dirt or moisture and will prevent its subsequent ingress.

If a cable oiler is not available, an alternative is to remove the cable from the machine. Hang the cable upright and make up a small funnel arrangement using plasticene or by taping a plastic bag around the upper end. Fill the funnel with oil and leave it overnight to drain through. Note that where nylon-lined cables are fitted, they should be used dry or lubricated with a silicone-based lubricant suitable for this application. On no account use ordinary engine oil because this will cause the liner to swell, pinching the cable.

Check all pivots and control levers, cleaning and lubricating them to prevent wear or corrosion. Where necessary, dismantle and clean any moving part which may have become stiff in operation.

3 *Final drive chain: cleaning, lubrication and adjustment*

In order that final drive chain life can be extended as much as possible, regular lubrication and adjustment is essential. This is particularly so when the chain is not enclosed or is fitted to a machine transmitting high power to the rear wheel. The chain may be lubricated whilst it is in place on the machine by the application of one of the proprietary chain greases contained in an aerosol can. Ordinary engine oil can be used, though owing to the speed with which it is flung off the rotating chain, its effective life is limited.

Accumulations of old lubricant and road dirt should be removed prior to the application of chain grease. This can be done by cleaning the chain with paraffin and a stiff brush,

Check tyre pressures using an accurate gauge

Add engine oil via filler hole as shown

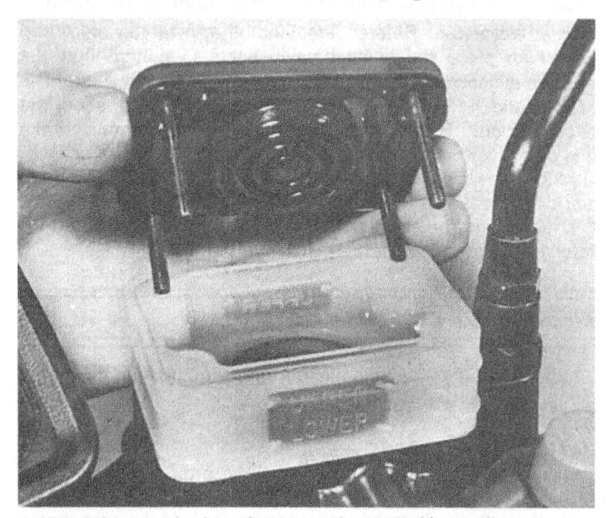

Brake fluid must be kept between upper and lower lines

Electrolyte can be checked via translucent case

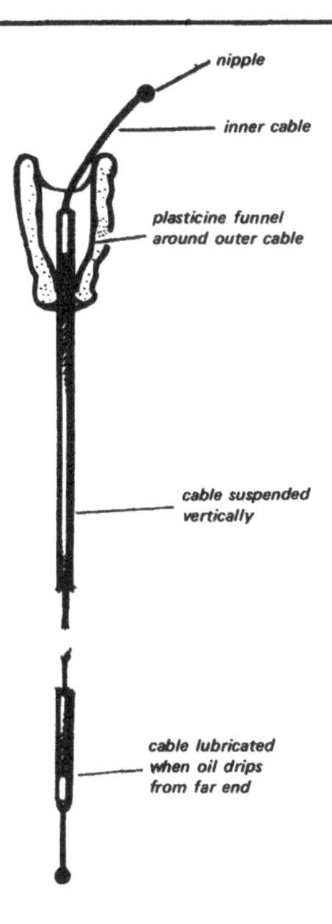

Oiling a control cable

rinsing with petrol to remove the dirt and paraffin. The petrol will evaporate, leaving the chain clean and dry. Lubrication can now be carried out. It will be noted that the standard chain is of the endless type, which effectively precludes chain removal prior to cleaning.

Check that the drive chain slack is correct by placing the machine on its centre stand and measuring the amount of vertical free play at the centre of the lower run. Chains rarely wear or stretch evenly, so rotate the rear wheel until the position is found at which the chain is tightest. The correct play is 15 – 25 mm ($\frac{5}{8}$ – 1 in).

If adjustment is required, remove the split pin from the castellated wheel spindle nut. Slacken the wheel spindle nut and the chain tensioner locknuts. Moving **both** adjuster bolts by an equal amount, set the chain to the prescribed tension. Note that alignment marks are provided on the adjusters. These must be equidistant from the end of the swinging arm on both sides to preserve wheel alignment. When correctly tensioned, secure the locknuts, tighten the wheel spindle nut and secure it with a new split pin.

Clean chain and use aerosol lubricant

Two monthly or every 1800 miles (3000 km)

Complete all the tasks listed under the previous mileage/time headings and then carry out the following:

1 Changing the engine/gearbox oil

Because the engine/gearbox oil is cleaned only by a gauze filter and not a full flow micro-pore element oil changes must be carried out on a regular basis at relatively short intervals if oil contamination and thus engine wear is to be kept to a minimum.

The oil should be drained when the engine is hot because the oil is thinner and will therefore drain more easily and more completely. Place a container of more than 1.7 lit (3 pints) below the crankcase. The drain plug is situated at the left-hand side of the casing. Remove the plug and also the filler plug/dipstick and allow all the oil to drain. Check the condition of the sealing washer before refitting and tightening the drain plug.

Replenish the engine with 1.7 lit (3 pints) SAE10W/40 engine oil. Allow the oil to settle and then check that the oil level comes within the upper/lower level lines on the dipstick. Allow the filler plug/dipstick to rest on the filler hole edge when checking the oil, do not screw it in.

Four monthly, or every 3600 miles (6000 km)

Complete the operations listed under the previous service headings, then carry out the following:

1 Cleaning the air filter element

It is vitally important that the air filter element is kept clean and in good condition if the engine is to function properly. If the element becomes choked with dust it follows that the airflow to the engine will be impaired, leading to poor performance and high fuel consumption. Conversely, a damaged filter will allow excessive amounts of unfiltered air to enter the engine, which can result in an increased rate of wear and possibly damage due to the weak nature of the mixture. The interval specified above indicates the maximum time limit between each cleaning operation. Where the machine is used in particularly adverse conditions it is advised that cleaning takes place on a much more frequent basis.

Tensioners have alignment marks

The element is retained in the air cleaner casing, access to which is via the left-hand side panel. This should be pulled off to reveal the air filter cover which is retained by three screws. Release the filter assembly from the air cleaner casing and peel off the foam filter band from its supporting framework.

The foam can be cleaned by washing in a high flash point solvent, such as white spirit. The use of petrol (gasoline) is not approved by the manufacturer in view of the potential fire risk. Allow the element to dry, then impregnate the foam with SAE 80 or 90 gear oil, removing any excess by squeezing it out. The element can now be reassembled and fitted.

If inspection has revealed any holes or tears, the element must be renewed immediately. On no account be tempted to omit the element in view of the damage that may ensue from the resulting weak mixture.

2 Checking the fuel pipe condition

Give the pipe which connects the fuel tap and carburettor a close visual examination, checking for cracks or any signs of leakage. In time, the synthetic rubber pipe will tend to deteriorate, and will eventually leak. Apart from the obvious fire risk, the evaporating fuel will affect fuel economy. If the pipe is to be renewed, always use the correct replacement type to ensure a good leak-proof fit. Never use natural rubber tubing because this will tend to break up when in contact with petrol, and will obstruct the carburettor jets.

3 Cleaning and resetting the sparking plug

Detach the sparking plug cap, and using the correct spanner remove the sparking plug. Clean the electrodes using a wire brush followed by a strip of fine emery cloth or paper. Check the plug gap with a feeler gauge, adjusting it if necessary to within the range of 0.6 – 0.7 mm (0.024 – 0.028 in). Make adjustments by bending the outer electrode, never the inner (central) electrode.

Before fitting the sparking plug smear the threads with a graphited grease; this will aid subsequent removal.

4 Checking and adjusting the valve clearances

The accurate setting of valve clearances is essential if the engine is to function properly. If the clearance becomes too great, the valves will not open fully. This restricts the amount of fuel/air mixture entering the cylinder which in turn lessens the power produced on each firing stroke. The result is a noisy and inefficient engine. Conversely, too small a clearance will mean that the valves do not close fully, leading to a marked fall off in performance. More significantly, the escaping hot gases will quickly destroy the valve faces.

The valve clearances are set with the engine cold, which is best interpreted as after the machine has stood overnight, ensuring that it has cooled fully since it was last run. If the prescribed clearance is set at this engine temperature (below 95°F, 35°C) it will ensure that the valves open fully and close fully when the engine is at normal running temperature.

Start by removing the seat, which is secured by a bolt on each side. Lift the seat clear, then remove the single fixing bolt at the rear of the fuel tank. Check that the fuel tap is off and pull off the fuel feed pipe. The tank can now be pulled rearwards and removed.

The engine must be at the top dead centre (TDC) position on the compression stroke, and to this end the alternator rotor has a 'T' (timing) mark which should be aligned with the fixed index mark on the cover. Remove the two inspection caps from the left-hand outer cover and release the valve inspection covers. Remove the sparking plug so that the crankshaft can be rotated easily. Using a socket passed through the central inspection hole in the left-hand outer cover, turn the engine anti-clockwise until the scribed 'T' mark appears in the upper hole and is aligned with the index mark.

Check that the engine is at TDC on the compression stroke by ensuring that none of the valves are open. If any valves are not closed, turn the engine through another 360° and re-align the 'T' mark. At this position there should be detectable clearance between the valve stems and rockers. Measure this clearance using successive feeler gauges until the gap is known. The gauge should be a light sliding fit between the valve and rocker. Note the clearance of each valve, and if necesary adjust to the correct clearance which is shown below. Adjustment is carried out after the locknut has been slackened. Turn the square-headed adjuster until the required clearance is obtained, then hold the adjuster position whilst the locknut is secured. If required, a Honda dealer can supply a special tool to hold the small square adjuster head, although most owners will be able to improvise with a small open-ended or adjustable spanner. After adjustment, turn the crankshaft through several revolutions and re-check the setting before refitting the covers, caps, tank and seat.

Valve clearances (with cold engine)

Inlet	*0.05 mm (0.002 in)*
Exhaust	*0.10 mm (0.004 in)*

Note that the decompressor cable adjustment must be checked whenever the valve clearances are checked or adjusted. The procedure is described below.

5 Adjusting the decompressor cable

The Honda CB250 RS model is equipped with an automatic decompressor, or valve lifter, as an aid to starting. The decompressor is controlled by a cam arrangement on the kickstart shaft and operates via an adjustable cable. The decompressor lever operates on the exhaust valves, holding them open to allow easy cranking of the engine. If the valve clearances are altered for any reason it is necessary to check that the decompressor lever still has the necessary amount of free play. Incorrect adjustment could result in the system not working correctly, or if over-tight, burnt exhaust valves.

With the alternator rotor 'T' mark aligned as described in Section 4 and the engine on the compression stroke, check the clearance in the decompressor mechanism, measured at the cylinder head lever. When correctly adjusted there should be 1 – 3 mm (0.04 – 0.12 in) free play measured at the lever end. If necessary, slacken the adjuster locknut and turn the adjuster to obtain the correct clearance. Do not omit to secure the locknut on completion of adjustment.

6 Checking the carburettor settings

Check that both of the throttle cables are clean and undamaged, ensuring that they are correctly routed and have not become kinked or trapped at any point. The cable adjustment should be set so that there is 2 – 6 mm (0.08 – 0.24 in) free play measured at the flanged inner edge of the rubber twistgrip. Adjustment is effected by means of an adjuster at the lower end of the cable. Further fine adjustment can be made at the upper adjuster.

Check the choke cable adjustment. With the choke control pulled out to its fullest extent check that the choke butterfly is fully closed; this can be detected by feeling for further movement of the operating lever on the carburettor. If adjustment is required this can be made at the cable securing clamp on the carburettor body.

The manufacturer recommends that the idle speed is checked and set to the prescribed 1200 ± 100 rpm, using the large knurled throttle stop control at the top of the instrument.

Having checked the throttle and choke cable adjustment and set the idle speed as described previously, proceed to check the fast idle speed. Run the engine until fully warmed up and then with the engine idling and the transmission in neutral, pull the choke control knob out to its fullest extent and note the tachometer reading. This should be within the specified range of 2000 – 2500 rpm. If adjustment is necessary, stop the engine and slacken the locknut securing the adjuster screw at the base of the fast idle link. Adjust the screw by the required amount, tighten its locknut and recheck the fast idle speed.

The pilot mixture is set with the engine at normal operating temperature, and where necessary the machine should be taken for a short ride to achieve this. With the engine idling at the prescribed speed, screw the pilot screw inwards (clockwise) until the engine begins to slow and falter. The pilot screw is located on the underside of the carburettor and passes upwards into the main body. Noting the number of turns or part turns of the screw, turn it anti-clockwise until it again begins to falter. The correct setting is midway between these two extremes, normally about $1\frac{3}{4}$ turns out from the fully closed position. After adjustment is complete, re-check the idle speed which may have risen slightly. If necessary, readjust it to the correct 1200 rpm setting.

7 *Brake inspection and adjustment*

Front brake

Check the entire hydraulic system for any sign of leakage or damage to the hoses or pipes, referring to Chapter 5 for further information should attention to these components be necessary. If the hydraulic fluid level has dropped significantly check very carefully for signs of fluid leakage at all unions and around the brake caliper. Note that the fluid level will fall slightly as the pads wear down, but should not change suddenly.

Using the inspection window provided in the front brake caliper, check the condition of the front brake pads. A red line is engraved around the friction material and indicates the extent of allowable wear. If either pad is worn down to its red line the pair should be renewed (see Chapter 5, Section 5).

Rear brake

Before any rear brake adjustment is made, check that the pedal is set at a suitable height. If necessary, make any adjustment required by altering the position of the pedal stop bolt. Note that any adjustment of the pedal height will have some effect on brake adjustment.

The brake should be set so that the end of the pedal is free to move 20 – 30 mm (0.8 – 1.2 in) before the brake begins to operate. Adjustment is effected by turning the nut at the end of the brake operating rod.

Check that when the brake is fully applied, the pointer on the brake actuating lever does not move beyond the limit of the wear indicator on the brake backplate. If it shows excessive wear it will be necessary to check and possibly renew the brake shoes. Refer to Chapter 5 for details.

When adjustment is complete, check that the brake light comes on as the pedal is depressed. If required, turn the brake switch adjusting nut to set the switch in the right position.

Remove air cleaner assembly, clean and re-oil foam

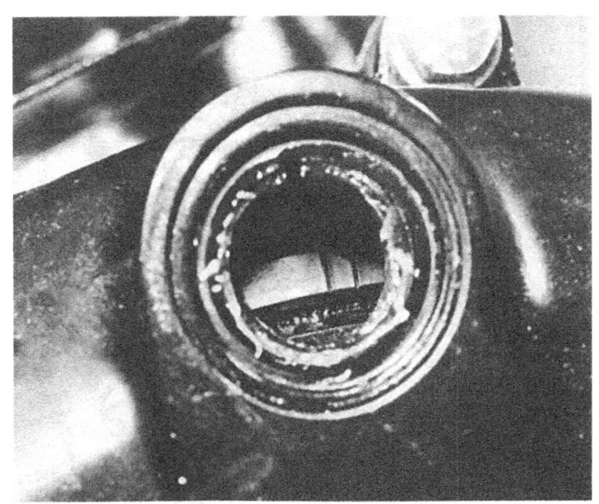

"T" mark must be aligned as shown

Check and set valve clearances as shown

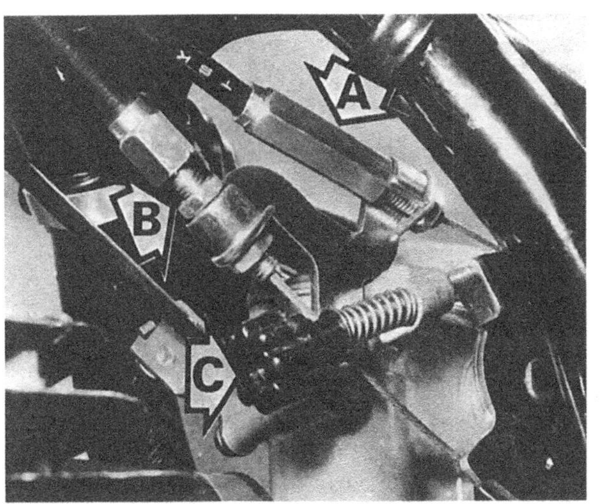

A: Closing cable B: Opening cable C: Throttle stop screw

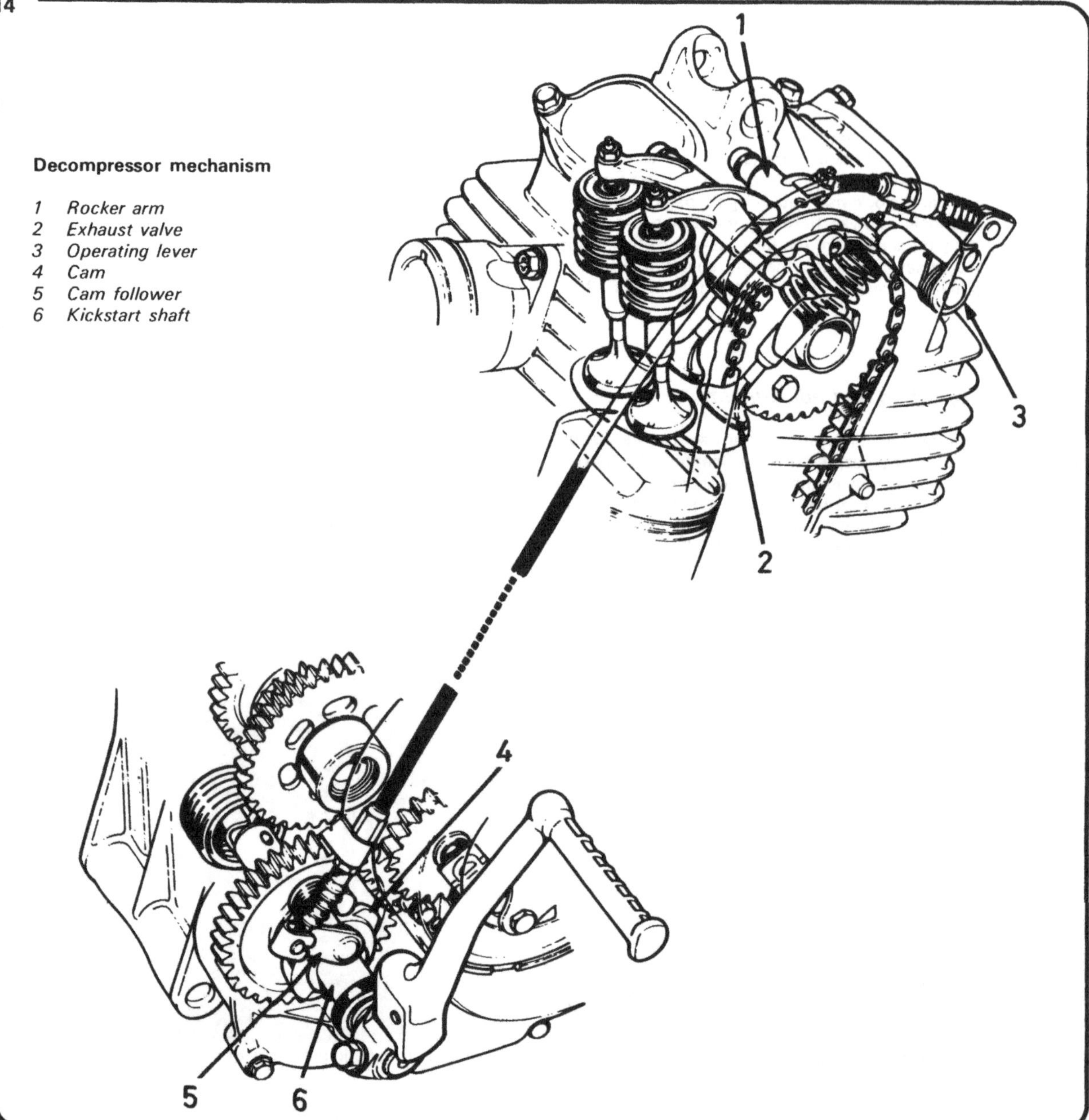

Pad wear can be checked via window in caliper

Renew pads when worn down to engraved line (arrowed)

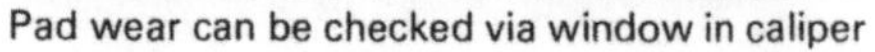

Pedal height can be set using stop bolt and locknut

Mark on brake plate indicates brake wear limit

8 Check clutch operation and adjustment

Clutch adjustment will be necessary to compensate for wear in the clutch plates and should be carried out at the prescribed interval, or whenever excessive lever travel is evident. The manufacturer recommends that the clutch is adjusted to give 10 – 20 mm (0.4 – 0.8 in) free play at the lever end. If significant adjustment is required it is best to make it at the lower adjuster, the handlebar lever adjuster can then be used for fine adjustment.

If the clutch becomes stiff or jerky in operation, check that the cable is clean, well lubricated and undamaged. Note that a stiff cable can add greatly to the lever pressure required to operate the clutch. Further details on clutch dismantling and overhaul will be found in Chapter 1.

9 Checking the front and rear suspension

Ensure that the front forks operate smoothly and progressively by pumping them up and down whilst the front brake is held on. Any faults revealed by this check should be investigated further, because any deterioration in the handling of the machine can have serious consequences if left unremedied. Check the condition of the fork stanchions. As with most current production machines, the fork stanchions are left exposed in the interests of fashion, and are thus prone to damage from stone chips or abrasion. Any damage to the stanchions will lead to rapid wear of the fork seals and can only be cured by renewing the stanchions. This is both costly and time consuming, so it is worth checking that the area below each dust seal is kept clean and greased. Remove any abrasive grit which may have accumulated around the dust seal lip. The above problems can be eliminated by fitting fork gaiters, these being available from most accessory stockists.

The rear suspension can be checked with the machine on the centre stand. Check that all of the suspension components are securely attached to the frame. Check for free play in the swinging arm by pushing and pulling it horizontally. Assuming that all is well, complete the operation by greasing the swinging arm pivot via the grease nipples provided on each side of the swinging arm.

10 General checks and lubrication

Check around the machine, looking for loose nuts, bolts or screws, retightening them as necessary. Check the stand and lever pivots for security and lubricate them with light machine oil or engine oil. Make sure that the stand springs are in good condition.

Swinging arm pivots should be lubricated via grease nipples

11 Headlamp aim

The alignment of the headlamp should be checked periodically to ensure that the maximum amount of light is directed onto the road ahead rather than into the eyes of uncoming road users. In the UK, it is illegal to use a motor vehicle with an improperly adjusted headlamp, and similar regulations apply in most other countries. For a full description of the adjustment procedure, refer to Chapter 6.

12 Checking wheel condition

Place the machine on the centre stand so that the front wheel is raised clear of the ground. Spin the wheel and check the rim alignment. Small irregularities can be corrected by tightening the spokes in the affected area although a certain amount of experience is necessary to prevent over-correction. Any flats in the wheel rim will be evident at the same time. These are more difficult to remove and in most cases it will be necessary to have the wheel rebuilt on a new rim. Apart from the effect on stability, a flat will expose the tyre bead and walls to greater risk of damage if the machine is run with a deformed wheel.

Check for loose and broken spokes. Tapping the spokes is the best guide to tension. A loose spoke will produce a quite different sound and should be tightened by turning the nipple in an anti-clockwise direction. Always check for run out by spinning the wheel again. If the spokes have to be tightened by an excessive amount, it is advisable to remove the tyre and tube as detailed in Chapter 5. This will enable the protruding ends of the spokes to be ground off, thus preventing them from chafing the inner tube and causing punctures. The condition of the rear wheel can be checked in exactly the same way as described above.

Eight monthly or every 7200 miles (12 000 km)

The following operations should be completed in addition to those specified under the previous headings:

1 *Cleaning the oil filter screen*

Oil is drawn into the lubrication system via a gauze filter screen located in the right-hand crankcase half. As this is the only form of oil filtration, it is essential that the screen is removed for cleaning on a regular basis. It will be appreciated that it is advantageous to carry out this operation in conjunction with an oil change, because it is necessary to drain the engine transmission oil before the screen can be removed.

Having drained the oil, release the kickstart lever after removing its pinch bolt, and remove the rear brake pedal and stop plate. The clutch and decompressor cables may be left connected provided that the casing is lodged within easy reach of the cables after removal. Remove the hexagon-headed screws which retain the cover and lift it away, noting that some provision must be made to catch any residual oil which may be released.

The rectangular gauze screen is located in a horizontal slot in the crankcase and may be pulled from position for cleaning. Wash the gauze in clean petrol, then slide it back into its casing slot. Before the cover is fitted, check the balancer adjustment as described in Section 3 below. Note that care should be exercised when fitting the cover to ensure that the decompressor mechanism engages correctly on the kickstart shaft assembly. If necessary, temporarily release the decompressor cable to facilitate engagement, remembering to adjust the cable after the cover is in place. (See Section 5, Four monthly/3600 miles for details.)

2 *Renew the sparking plug*

The manufacturer recommends that the sparking plug is renewed as a precautionary measure at this stage. Always ensure that a plug of the correct type and heat range is fitted, and that the gap is set to the prescribed 0.6 – 0.7 mm (0.024 – 0.028 in) prior to installation. If the old plug is in reasonable condition, it can be cleaned and re-gapped and carried as an emergency spare in the toolbox.

3 *Checking and adjusting the balancer chain*

The engine unit is fitted with a pair of balancer weights which counteract the normal out-of-balance forces found in a single-cylinder engine. The rear balancer weight runs on the left-hand end of the gearbox mainshaft, whilst the front weight is mounted on the right-hand end of a separate balancer shaft. The two are connected and timed by a single-row chain driven from the left-hand end of the crankshaft.

To facilitate adjustment to compensate for chain wear, the front balancer shaft runs in the eccentric bore of a tubular holder. The right-hand end of the holder terminates in an adjuster plate which is locked by a single retaining bolt. When adjustment is necessary the holder is rotated in the crankcase, effectively moving the axis of the front balancer shaft forwards or backwards.

Balancer chain adjustment requires the removal of the right-hand outer casing, and it is thus convenient to carry out this operation whilst the oil filter screen is being dealt with. Identify the balancer holder lock bolt, which passes through the adjuster plate's elongated lock bolt hole. Slacken and remove the bolt, noting that the holder should rotate anti-clockwise under spring pressure. Make sure that the holder rotates freely by pulling it back against spring tension and releasing it. This should result in its springing back until the balancer chain is taut.

The lower edge of the adjuster plate is marked by a series of lines. The balancer chain tension is set by moving the adjuster plate back (clockwise) by one graduation from the fully taut position. Holding this position refit and tighten the locking bolt to 2.2 – 2.8 kgf m (16 – 20 lbf ft).

Occasionally it may prove impossible to obtain sufficient adjustment within the range of movement provided by the elongated slot. If this proves to be the case it will be necessary to move the adjuster plate in relation to the holder. Disconnect the adjuster spring. Release the circlip which retains the balance weight, then slide the latter off the end of the balancer shaft. Remove the plain washer which is fitted behind the balancer weight, then release the large circlip which secures the adjuster plate to the holder. It will be seen that the adjuster plate is located by tangs which engage in corresponding slots in the holder. Withdraw the adjuster plate and reposition it one slot further round (clockwise) to bring the range of adjustment within the scope of the plate.

Reassemble the balancer components by reversing the dismantling sequence, noting that the balancer timing mark must align with its counterpart on the balancer shaft. The tensioning operation can now be completed as described above.

Slacken lock bolt to permit balancer chain adjustment

4 *Checking the steering head bearings*

Wear or play in the steering head bearings will cause imprecise handling and can be dangerous if allowed to develop unchecked. Test for play by pushing and pulling on the handlebars whilst holding the front brake on. Any wear in the head races will be seen as movement between the fork yokes and the steering lug.

Before carrying out adjustment, place a wooden crate or similar item beneath the crankcase so that the front wheel is raised clear of the ground. Check that the handlebars will turn smoothly and freely from lock to lock. If the steering feels notchy or jerky in operation it may be due to worn or damaged bearings. Should this be suspected it will be necessary to overhaul the steering head bearings as described in Chapter 4.

To adjust the steering head bearings, slacken the large steering stem nut at the centre of the top fork yoke, then use a C-spanner to tighten the slotted adjuster nut immediately below the top yoke. As a guide to adjustment, tighten the slotted nut until a firm resistance is felt, then back it off by $\frac{1}{8}$ turn. The object is to remove all discernible play without applying any appreciable preload. It should be noted that it is possible to apply a loading of several tons on the small steering head bearings without this being obvious when turning the handlebars. This will cause an accelerated rate of wear, and thus must be avoided.

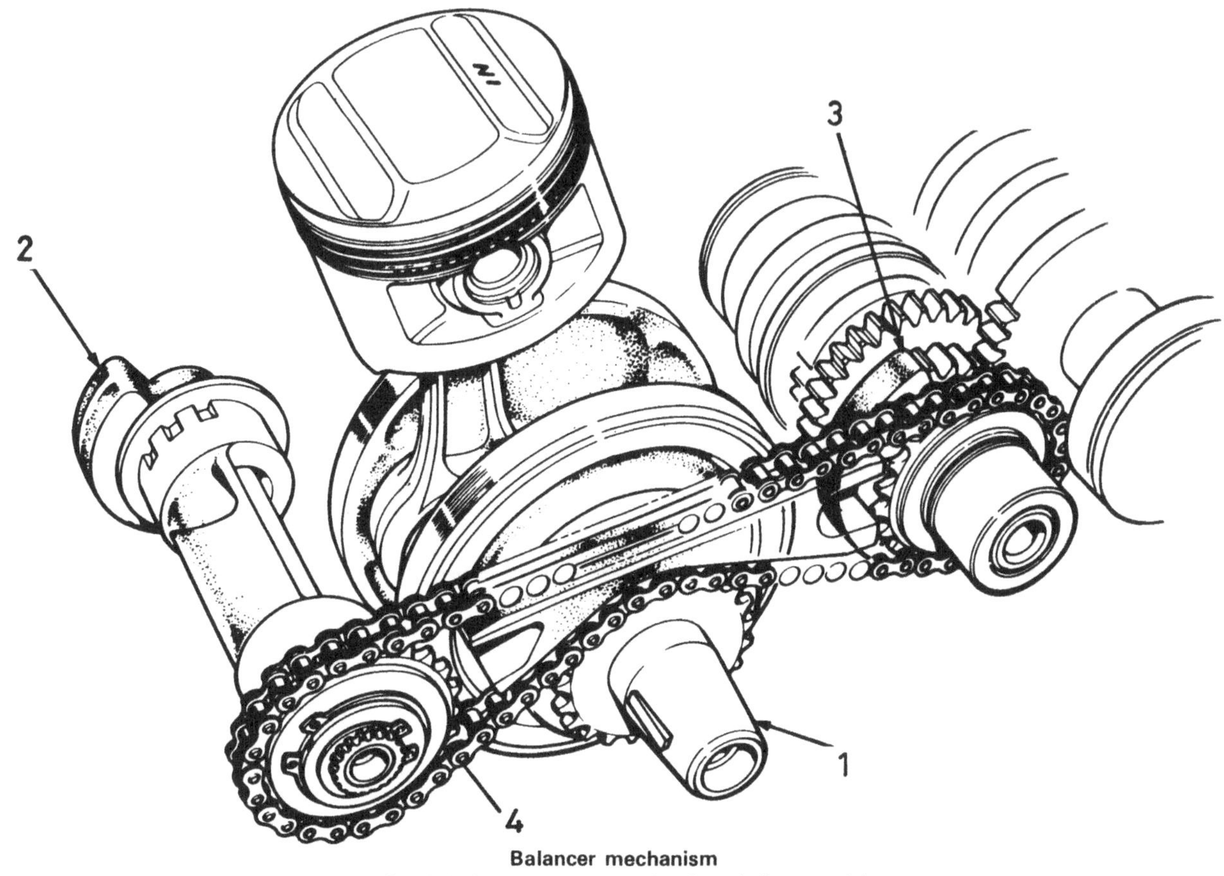

Balancer mechanism

1 Crankshaft
2 Front balance weight
3 Rear balance weight
4 Balancer chain

Yearly or every 10 800 miles (18 000 km)

Complete all the tasks listed under the previous time/mileage intervals and then complete the following:

1 Changing the brake fluid

Honda recommend that the hydraulic fluid used in the braking system is changed at 10 800 mile intervals as a precautionary measure. Where the machine covers less than the above mileage per year the fluid should be changed at **yearly** intervals. Hydraulic fluid is hygroscopic, which means that moisture in the air will gradually be absorbed by the fluid. Although the system is effectively sealed, the fluid will gradually deteriorate and thus must be renewed before the contamination lowers its boiling point to an unsafe level.

Before commencing work, obtain a can of DOT 3 or SAE J1703 hydraulic fluid, a length of clear plastic tubing which is a tight fit on the caliper bleed nipple and a clean glass jar. Remove the rubber dust cap from the bleed nipple and attach the plastic tubing. Place the free end of the tube in the jar.

Remove the master cylinder reservoir cap, open the bleed valve and operate the lever (or pedal) until all of the old hydraulic fluid has been expelled. Where twin discs are fitted, close the bleed valve, connect the tubing to the remaining caliper and repeat the operation.

Fill the reservoir with clean brake fluid and operate the lever or pedal until the system is filled, making sure that the fluid in the reservoir is kept above the minimum level. To remove any residual air from the system, close the bleed valve and operate the lever until it feels hard. Keeping pressure on the lever or pedal slacken the bleed valve to allow the fluid and any air to be expelled, then close the valve. Repeat this sequence until it is certain that no further air bubbles remain. When bleeding has been completed refit the dust cap on the bleed nipple, top up the reservoir to the upper level line and refit the reservoir cap. Always discard all used fluid immediately.

Additional routine maintenance

Certain aspects of routine maintenance make it impossible to place operations under specific mileage or calendar headings, or may require modification of the latter. A good example is the effect of a predominantly dusty environment on certain maintenance operations. In this case, the air cleaner element and chain maintenance intervals can be reduced considerably to prevent the clogging of the former and accelerated wear of the latter. Similar advice can be applied to general maintenance and lubrication of the various controls, pivots and cables. Some seasonal variations may be applicable, particularly where the drive chain and other exposed areas are vulnerable to rain, snow and road salting.

In addition to the above environmental considerations, some components will deteriorate at rates dependent on usage rather than mileage or age. This will affect many areas, but a significant example is shown below.

1 Side stand – pad renewal

The side stand is fitted with a rubber pad which will gradually wear down with use. Check the pad condition and renew it when it nears the raised wear line. The old pad can be released by removing the single retaining bolt and the new item fitted by reversing the dismantling sequence.

Pad should be renewed when worn to line (arrowed)

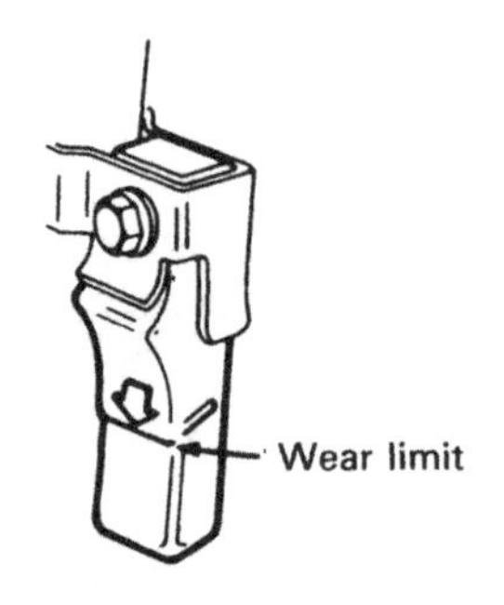

Side stand pad wear limit

Quick glance maintenance adjustments and capacities

Engine/gearbox oil capacity		
Dry	2 lit (3.6 Imp pint)	
At oil change	1.7 lit (3.0 Imp pint)	
Sparking plug type	NGK DR8ES or ND X27ESR-U	
Sparking plug gap	0.6 – 0.7 mm (0.024 – 0.028 in)	
Valve clearances (cold)		
Inlet	0.05 mm (0.002 in)	
Exhaust	0.10 mm (0.004 in)	
Front fork oil capacity		
Per leg	155 – 160 cc (5.24 – 5.40 Imp fl oz)	
Tyre pressures	**Front**	**Rear**
Solo	24 psi (1.75 kg/cm^2)	32 psi (2.25 kg/cm^2)
With pillion	24 psi (1.75 kg/cm^2)	36 psi (2.50 kg/cm^2)

Recommended lubricants

Components	Lubricant
Engine/transmission	
General, all-temperature use	SAE 10W/40
Above 15°C (60°F)	SAE 30
-10° to +15°C (15° – 60°F)	SAE 20 or 20W
Above -10°C (15°F)	SAE 20W/50
Below 0°C (32°F)	SAE 10W
Front forks	Automatic transmission fluid (ATF) or fork oil
Chain	Aerosol chain lubricant
General lubrication	Light machine oil
Wheel bearings	High melting point grease
Swinging arm	High melting point grease
Disc brake	DOT 3 or SAE J1703 brake fluid

HONDA CB250 RS

Check list

Weekly or every 200 miles (320 km)

1. Check the tyre pressures
2. Check the engine/gearbox oil level
3. Check the brake fluid level in the front brake reservoir

Monthly or every 600 miles (1000 km)

1. Check the battery electrolyte level
2. Lubricate the controls, cables and pivot points
3. Clean, lubricate and adjust the rear chain

Two monthly or every 1800 miles (3000 km)

1. Change the engine/gearbox oil

Four monthly or every 3600 miles (6000 km)

1. Clean and lubricate the air filter element
2. Check the condition of the fuel pipe
3. Clean and reset the sparking plug
4. Check and adjust the valve clearances
5. Adjust the decompressor cable free play
6. Check the carburettor settings
7. Inspect the condition of both brake systems
8. Check and adjust the clutch operating mechanism
9. Check the front and rear suspension for correct operation
10. Make a general safety check of fasteners
11. Check and adjust the headlamp beam alignment
12. Check the condition of both wheels
13. Clean the fuel float bowl filter gauze – CB250 RSD-C

Eight monthly or every 7200 miles (12 000 km)

1. Remove and clean the engine oil filter screen
2. Renew the sparking plug
3. Adjust the balance chain tension
4. Check and if necessary adjust the steering head bearings

Yearly or every 10 800 miles (18 000 km)

1. Change the front brake fluid

Adjustment data

Tyre pressures	**Front**	**Rear**
Solo	24 psi (1.75 kg/cm²)	32 psi (2.25 kg/cm²)
With pillion	24 psi (1.75 kg/cm²)	36 psi (2.50 kg/cm²)

Sparking plug type NGK DR8ES or ND X27ESR-U

Sparking plug gap 0.6 – 0.7 mm (0.024 – 0.028 in)

Inlet 0.05 mm (0.002 in)
Exhaust 0.10 mm (0.004 in)

Ignition timing

CB250 RS-A
Initial 12° BTDC @ 1200 rpm
Full advance 25°BTDC @ 3450 rpm

CB250 RSD-C
Initial 15° BTDC @ 1200 rpm
Full advance 37 ± 2° BTDC @ 3450 rpm

Idle speed 1200 ± 100 rpm

Fast idle speed 2000 – 2500 rpm

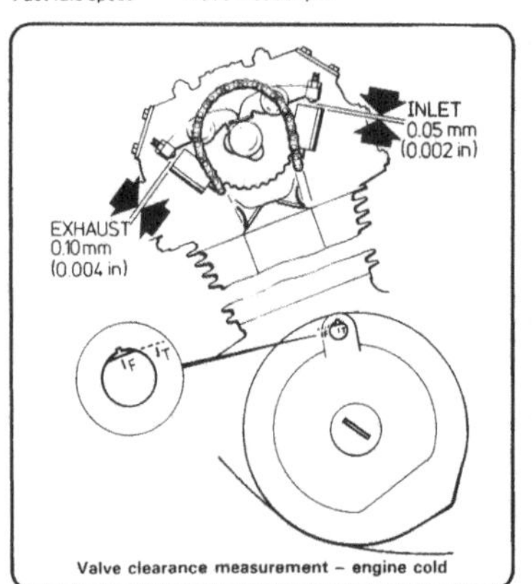

Valve clearance measurement – engine cold

Recommended lubricants

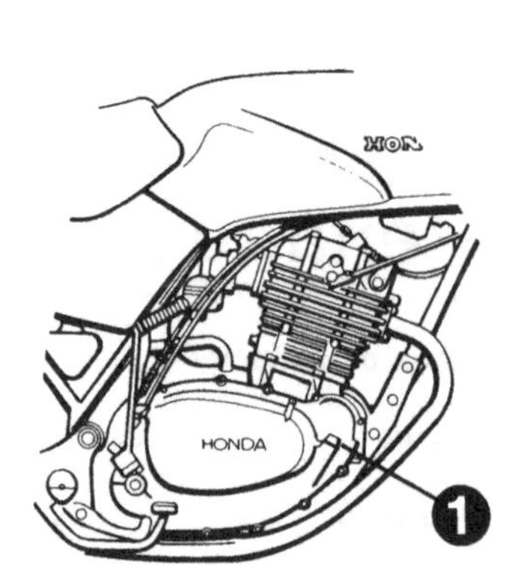

Component	Quantity	Type/viscosity
1 Engine/gearbox	1.7 lit (3.0 Imp pint)	SAE 10W/30 engine oil
2 Front forks	155 – 160cc	ATF or fork oil
3 Chain	As required	Aerosol chain lubricant
4 Wheel bearings	As required	High melting-point grease
5 Swinging arm	As required	High melting-point grease
6 Pivot points	As required	Multi-purpose grease
7 Control cables	As required	Light oil
8 Disc brake	As required	DOT 3 or SAE J1703 hydraulic brake fluid

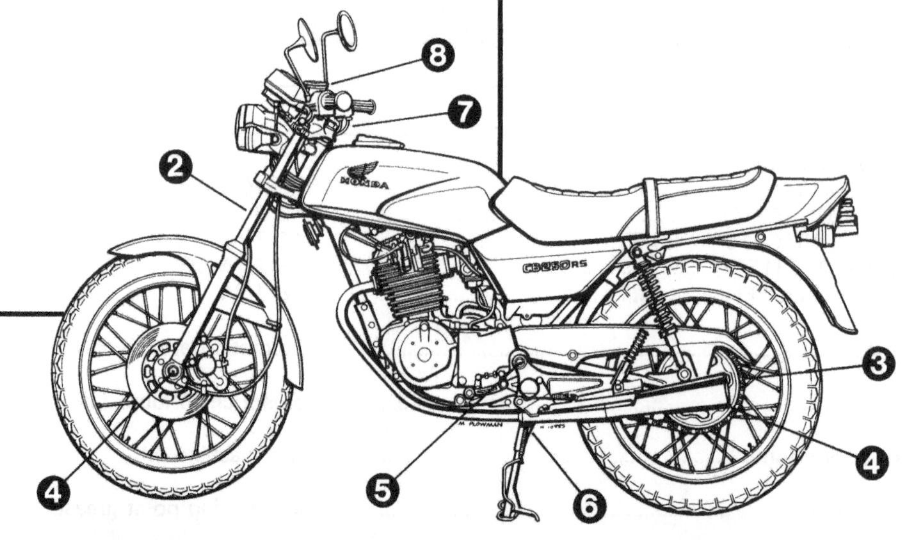

ROUTINE MAINTENANCE GUIDE

Working conditions and tools

When a major overhaul is contemplated, it is important that a clean, well-lit working space is available, equipped with a workbench and vice, and with space for laying out or storing the dismantled assemblies in an orderly manner where they are unlikely to be disturbed. The use of a good workshop will give the satisfaction of work done in comfort and without haste, where there is little chance of the machine being dismantled and reassembled in anything other than clean surroundings. Unfortunately, these ideal working conditions are not always practicable and under these latter circumstances when improvisation is called for, extra care and time will be needed.

The other essential requirement is a comprehensive set of good quality tools. Quality is of prime importance since cheap tools will prove expensive in the long run if they slip or break when in use, causing personal injury or expensive damage to the component being worked on. A good quality tool will last a long time, and more than justify the cost.

For practically all tools, a tool factor is the best source since he will have a very comprehensive range compared with the average garage or accessory shop. Having said that, accessory shops often offer excellent quality tools at discount prices, so it pays to shop around. There are plenty of tools around at reasonable prices, but always aim to purchase items which meet the relevant national safety standards. If in doubt, seek the advice of the shop proprietor or manager before making a purchase.

The basis of any tool kit is a set of open-ended spanners, which can be used on almost any part of the machine to which there is reasonable access. A set of ring spanners makes a useful addition, since they can be used on nuts that are very tight or where access is restricted. Where the cost has to be kept within reasonable bounds, a compromise can be effected with a set of combination spanners – open-ended at one end and having a ring of the same size on the other end. Socket spanners may also be considered a good investment, a basic 3/8 in or 1/2 in drive kit comprising a ratchet handle and a small number of socket heads, if money is limited. Additional sockets can be purchased, as and when they are required. Provided they are slim in profile, sockets will reach nuts or bolts that are deeply recessed. When purchasing spanners of any kind, make sure the correct size standard is purchased. Almost all machines manufactured outside the UK and the USA have metric nuts and bolts, whilst those produced in Britain have BSF or BSW sizes. The standard used in USA is AF, which is also found on some of the later British machines. Others tools that should be included in the kit are a range of crosshead screwdrivers, a pair of pliers and a hammer.

When considering the purchase of tools, it should be remembered that by carrying out the work oneself, a large proportion of the normal repair cost, made up by labour charges, will be saved. The economy made on even a minor overhaul will go a long way towards the improvement of a toolkit.

In addition to the basic tool kit, certain additional tools can prove invaluable when they are close to hand, to help speed up a multitude of repetitive jobs. For example, an impact screwdriver will ease the removal of screws that have been tightened by a similar tool, during assembly, without a risk of damaging the screw heads. And, of course, it can be used again to retighten the screws, to ensure an oil or airtight seal results. Circlip pliers have their uses too, since gear pinions, shafts and similar components are frequently retained by circlips that are not too easily displaced by a screwdriver. There are two types of circlip pliers, one for internal and one for external circlips. They may also have straight or right-angled jaws.

One of the most useful of all tools is the torque wrench, a form of spanner that can be adjusted to slip when a measured amount of force is applied to any bolt or nut. Torque wrench settings are given in almost every modern workshop or service manual, where the extent to which a complex component, such as a cylinder head, can be tightened without fear of distortion or leakage. The tightening of bearing caps is yet another example. Overtightening will stretch or even break bolts, necessitating extra work to extract the broken portions.

As may be expected, the more sophisticated the machine, the greater is the number of tools likely to be required if it is to be kept in first class condition by the home mechanic. Unfortunately there are certain jobs which cannot be accomplished successfully without the correct equipment and although there is invariably a specialist who will undertake the work for a fee, the home mechanic will have to dig more deeply in his pocket for the purchase of similar equipment if he does not wish to employ the services of others. Here a word of caution is necessary, since some of these jobs are best left to the expert. Although an electrical multimeter of the AVO type will prove helpful in tracing electrical faults, in inexperienced hands it may irrevocably damage some of the electrical components if a test current is passed through them in the wrong direction. This can apply to the synchronisation of twin or multiple carburettors too, where a certain amount of expertise is needed when setting them up with vacuum gauges. These are, however, exceptions. Some instruments, such as a strobe lamp, are virtually essential when checking the timing of a machine powered by CDI ignition system. In short, do not purchase any of these special items unless you have the experience to use them correctly.

Although this manual shows how components can be removed and replaced without the use of special service tools (unless absolutely essential), it is worthwhile giving consideration to the purchase of the more commonly used tools if the machine is regarded as a long term purchase Whilst the alternative methods suggested will remove and replace parts without risk of damage, the use of the special tools recommended and sold by the manufacturer will invariably save time.

Standard torque settings

Specific torque settings will be found at the end of the specifications section of each chapter. Where no figure is given, bolts should be secured according to the table below.

Fastener type (thread diameter)	kgf m	lbf ft
5 mm bolt or nut	0.45 – 0.6	3.5 – 4.5
6 mm bolt or nut	0.8 – 1.2	6 – 9
8 mm bolt or nut	1.8 – 2.5	13 – 18
10 mm bolt or nut	3.0 – 4.0	22 – 29
12 mm bolt or nut	5.0 – 6.0	36 – 43
5 mm screw	0.35 – 0.5	2.5 – 3.6
6 mm screw	0.7 – 1.1	5 – 8
6 mm flange bolt	1.0 – 1.4	7 – 10
8 mm flange bolt	2.4 – 3.0	17 – 22
10 mm flange bolt	3.0 – 4.0	22 – 29

Chapter 1 Engine, clutch and gearbox

For information relating to the CB250 RSD-C model, refer to Chapter 7

Contents

Specifications

Engine

Type	Air cooled, single cylinder, four stroke
Bore	74 mm (2.91 in)
Stroke	57.8 mm (2.28 in)
Capacity	248 cc (15.1 cu in)
Compression ratio	9.3 : 1
Lubrication system	Wet sump, forced lubrication

Piston

Type	Forged aluminium alloy
OD at skirt	73.97 – 73.99 mm (2.912 – 2.913 in)
Service limit	73.88 mm (2.909 in)
Gudgeon pin bore ID	19.002 – 19.008 mm (0.7481 – 0.7483 in)
Service limit	19.08 mm (0.751 in)
Gudgeon pin OD	18.994 – 19.000 mm (0.7478 – 0.7480 in)
Service limit	19.96 mm (0.747 in)

Piston rings

End gap (installed):	
Top and 2nd	0.15 – 0.35 mm (0.006 – 0.010 in)
Service limit	0.5 mm (0.02 in)
Oil	0.2 – 0.9 mm (0.007 – 0.035 in)
Service limit	N/A

Ring to groove clearance:

Top and 2nd	0.015 – 0.045 mm (0.0006 – 0.0018 in)
Service limit	0.12 mm (0.006 in)
Oil	0.017 mm (0.0007 in)

Cylinder bore

Diameter	74.00 – 74.01 mm (2.913 – 2.914 in)
Service limits:	
Diameter	74.11 mm (2.918 in)
Taper	0.05 mm (0.002 in)
Ovality	0.05 mm (0.002 in)
Cylinder to piston clearance	0.01 – 0.04 mm (0.0006 – 0.0018 in)
Service limit	0.1 mm (0.004 in)

Crankshaft

Big-end bearing axial clearance	0.05 – 0.45 mm (0.0020 – 0.0177 in)
Service limit	0.6 mm (0.0236 in)
Big-end bearing radial clearance	0.006 – 0.018 mm (0.0002 – 0.0007 in)
Service limit	0.05 mm (0.002 in)
Small-end ID	19.020 – 19.041 mm (0.7488 – 0.7496 in)
Service limit	19.07 mm (0.7508 in)
Crankshaft runout	0.1 mm (0.0039 in)

Valves

Clearances (cold engine):		
Inlet	0.05 mm (0.002 in)	
Exhaust	0.10 mm (0.004 in)	
Timing	1 mm lift	(0 lift)
Inlet opens at	5° BTDC	(58° BTDC)
Inlet closes at	30° ABDC	(96° ABDC)
Exhaust opens at	35° BBDC	(83° BBDC)
Exhaust closes at	5° ATDC	(65° ATDC)
Valve stem OD:		
Inlet	5.475 – 5.490 mm (0.2037 – 0.2161 in)	
Service limit	5.465 mm (0.2152 in)	
Exhaust	5.455 – 5.470 mm (0.2148 – 0.2154 in)	
Service limit	5.445 mm (0.2144 in)	
Valve guide ID:		
Inlet	5.500 – 5.512 mm (0.2165 – 0.2170 in)	
Service limit	5.53 mm (0.218 in)	
Exhaust	5.500 – 5.512 mm (0.2165 – 0.2170 in)	
Service limit	5.53 mm (0.218 in)	
Stem to guide clearance:		
Inlet	0.010 – 0.047 mm (0.0004 – 0.0019 in)	
Service limit	0.06 mm (0.0024 in)	
Exhaust	0.030 – 0.057 mm (0.0012 – 0.0022 in)	
Service limit	0.07 mm (0.0028 in)	
Valve face width	1.2 – 1.4 mm (0.048 – 0.055 in)	
Service limit	2.0 mm (0.08 in)	
Valve seat width	1.2 – 1.4 mm (0.048 – 0.055 in)	
Service limit	2.0 mm (0.08 in)	
Valve spring free length:		
Inner	38.1 mm (1.50 in)	
Service limit	37.0 mm (1.46 in)	
Outer	36.24 mm (1.43 in)	
Service limit	35.3 mm (1.39 in)	
Spring preload at specified length:		
Inner	6.99 – 7.99 kg @ 26.0 mm (15.41 – 17.61 lb @ 1.024 in)	
Outer	12.39 – 13.99 kg @ 29.0 mm (27.31 – 30.84 lb @ 1.142 in)	

Rockers

Rocker arm ID	12.000 – 12.018 mm (0.4724 – 0.4731 in)
Service limit	12.05 mm (0.474 in)
Rocker shaft OD	11.966 – 11.984 mm (0.4711 – 0.4718 in)
Service limit	11.91 mm (0.469 in)

Cylinder head and camshaft

Maximum cylinder head warpage	0.1 mm (0.004 in)
Camshaft bearing ID:	
Left	20.000 – 20.021 mm (0.7874 – 7882 in)
Service limit	20.05 mm (0.789 in)
Right	24.000 – 24.021 mm (0.9449 – 0.9457 in)
Service limit	24.05 mm (0.947 in)

Camshaft journal OD:	
Left	19.954 – 19.975 mm (0.7856 – 0.7864 in)
Service limit	19.9 mm (0.78 in)
Right	23.954 – 23.975 mm (0.9431 – 0.9439 in)
Service limit	23.9 mm (0.94 in)

Balancer mechanism

Balancer holder OD	39.964 – 39.980 mm (1.5734 – 1.5740 in)
Service limit	39.91 mm (1.571 in)
Balancer holder ID	26.007 – 26.020 mm (1.0239 – 1.0244 in)
Service limit	26.05 mm (1.026 in)
Rear balance weight ID	26.007 – 26.020 mm (1.0239 – 1.0244 in)
Service limit	26.05 mm (1.026 in)

Clutch

Type	Wet, multiplate
Number of plates:	
Plain	4
Friction	5
Clutch springs	
Number	4
Free length	37.3 mm (1.46 in)
Service limit	35.8 mm (1.41 in)
Preload/length	23 ± 1.15 kg @ 23.5 mm (50.7 ± 2.5 lb @ 0.93 in)
Friction plate thickness	2.62 – 2.78 mm (0.102 – 0.109 in)
Service limit	2.3 mm (0.091 in)
Plain plate warpage (max)	0.3 mm (0.012 in)
Clutch drum bore	27.000 – 27.021 mm (1.0630 – 1.0638 in)
Service limit	27.05 mm (1.065 in)
Sleeve OD	26.959 – 26.980 mm (1.0614 – 1.0622 in)
Service limit	26.91 mm (1.059 in)
Sleeve ID	22.000 – 22.035 mm (0.8661 – 0.8675 in)
Service limit	22.05 mm (0.868 in)

Gearbox

Type	5 speed constant mesh
Primary reduction	2.464 : 1 (69/28)
Gear ratios:	
1st	2.800 : 1 (42/15)
2nd	1.850 : 1 (37/20)
3rd	1.375 : 1 (33/24)
4th	1.111 : 1 (30/27)
Top	0.931 : 1 (27/29)
Final reduction	3.142 : 1 (44/14)
Backlash between gear teeth (maximum)	0.2 mm (0.008 in)
Selector dog minimum clearance	0.3 mm (0.012 in)
Gear ID (Layshaft 1st and 3rd. Mainshaft (4th and 5th)	25.020 – 25.041 mm (0.9850 – 0.9859 in)
Service limit	25.10 mm (0.988 in)
Bushes:	
ID	20.020 – 20.041 mm (0.7866 – 0.7890 in)
Service limit	20.10 mm (0.791 in)
OD	25.005 – 25.016 mm (0.9844 – 0.9849 in)
Service limit	24.95 mm (0.982 in)
Gear to bush clearance	0.004 – 0.036 mm (0.0002 – 0.0014 in)
Service limit	0.15 mm (0.006 in)
Mainshaft OD	24.959 – 24.980 mm (0.9826 – 0.9835 in)
Service limit	24.91 mm (0.981 in)
Gear to shaft clearance	0.040 – 0.082 mm (0.0016 – 0.0032 in)
Service limit	0.15 mm (0.006 in)
Shaft to bush clearance	0.020 – 0.054 mm (0.0008 – 0.0021 in)
Service limit	0.15 mm (0.006 in)
Layshaft OD:	
At 3rd gear journal	24.959 – 24.980 mm (0.9826 – 0.9835 in)
Service limit	24.91 mm (0.981 in)
At 1st gear journal	19.987 – 20.000 mm (0.7869 – 0.7874 in)
Service limit	19.95 mm (0.785 in)
Selector fork ID:	
Centre fork	12.000 – 12.021 mm (0.4724 – 0.4733 in)
Service limit	12.05 mm (0.474 in)
Outer forks	15.000 – 15.021 mm (0.5906 – 0.5914 in)
Service limit	15.05 mm (0.593 in)

Claw thickness	4.93 – 5.00 mm (0.194 – 0.197 in)
Service limit	4.5 mm (0.18 in)
Selector fork shaft OD:	
Centre fork	11.966 – 11.984 mm (0.4711 – 0.4718 in)
Service limit	11.91 mm (0.469 in)
Outer forks	14.966 – 14.984 mm (0.5892 – 0.5899 in)
Service limit	14.91 mm (0.587 in)
Selector drum OD	11.966 – 11.984 mm (0.4711 – 0.4718 in)
Service limit	11.91 mm (0.469 in)
Selector drum bore ID	12.000 – 12.027 mm (0.4724 – 0.4735 in)
Service limit	12.10 mm (0.476 in)
Kickstart mechanism:	
Pinion ID	22.000 – 22.021 mm (0.8661 – 0.8670 in)
Service limit	22.10 mm (0.870 in)
Shaft OD	21.959 – 21.980 mm (0.8645 – 0.8654 in)
Service limit	21.91 mm (0.863 in)

Torque wrench settings

	kgf m	lbf ft
Cylinder barrel bolts	0.8-1.2	6-9
Cylinder head bolts	3.0-3.6	22-26
Cylinder head nuts	2.2-2.8	16-20
Upper crankcase bolts:		
6 mm	1.0-1.4	7-10
8 mm	2.0-2.6	14-19
Lower crankcase bolts:		
6 mm	1.0-1.4	7-10
8 mm	2.2-2.8	16-20
Clutch centre nut	4.5-6.0	33-43
Generator rotor nut	10.0-12.0	72-87
Automatic timing unit	4.5-6.0	33-43
Camshaft sprocket bolts	1.7-2.3	12-17
Oil drain plug	2.0-3.0	14-22

1 General description

The Honda CB250 RS employs an air-cooled four-stroke engine built in unit with the gearbox and clutch assemblies. The single-cylinder overhead camshaft (ohc) unit is very similar to those used in the XL and XR 250 models. The engine castings are of light alloy construction with a black painted finish. The cylinder barrel incorporates a steel liner.

The cylinder head is of the four valve type in the interests of low reciprocating weight and to permit a greater valve area per bore size than is possible with a two-valve design. A single carburettor feeds the bifurcated inlet passage, whilst two separate exhaust ports discharge into a twin pipe exhaust system.

The crankshaft is of conventional construction and is supported on journal ball main bearings, the big-end bearing being of the caged roller type. Primary drive is direct from the crankshaft pinion to the clutch outer drum. To counter the imbalance inherent in all single-cylinder engines a chain driven balancer system is employed. Two balance weights are used, the front one forming part of an assembly which permits balancer chain tension adjustment, whilst the rear weight is appended to the left-hand end of the gearbox mainshaft.

Engine power is transmitted through the clutch to the five-speed constant mesh gearbox, which is located at the rear of the main engine casings or crankcase halves. The engine, primary drive and gearbox share a common lubrication system in which oil from the crankcase sump is fed under pressure to the major engine and transmission components.

2 Operations with the engine/gearbox unit in the frame

The following items can be overhauled with the engine/gearbox unit installed in the frame.

a) Cylinder head and valves
b) Camshaft
c) Cylinder barrel and piston
d) Clutch and primary drive
e) Kickstart mechanism
f) Ignition pick-up
g) Gear selector mechanism
h) Front balance weight and sprocket
i) Alternator
j) Final drive sprocket

When several operations need to be undertaken simultaneously, it would probably be an advantage to remove the complete unit from the frame, a comparatively simple operation that should take approximately one hour. This will give the advantage of better access and more working space.

3 Operations with the engine/gearbox unit removed from the frame

It will be necessary to remove the engine/gearbox unit from the frame to gain access to the following:

a) Crankshaft assembly
b) Gearbox components
c) Rear balancer and balancer chain
d) Kickstart mechanism

4 Removing the engine/gearbox unit from the frame

1 Before commencing any dismantling work it will be necessary to drain the engine oil. This is best done whilst the engine is hot, so if time permits, drain the oil and leave the machine to cool down before starting work. The drain plug is located at the left-hand side of the crankcase to facilitate draining when the machine is supported by the side stand. The crankcase holds approximately 3.0 pints (1.7 litres) of oil, and a suitably sized drain tray should be used to collect the old oil.

2 Place the machine securely on its centre stand ensuring that it is in no danger of falling off during dismantling. It is a good precaution to place wooden blocks against the front wheel to prevent the machine from rolling forward. It is helpful, though by no means essential, to raise the machine a few feet from the

ground by placing it on a stout bench or similar support.
3 Remove the side panels by pulling them free of their locating rubbers. Slacken and remove the two seat retaining bolts. These are located on each side of the seat, passing through brackets below the seat base. Lift the seat away and place it to one side. Check that the fuel tap is turned to the Off position, then prise off the fuel pipe between the tap and the carburettor. Slacken the single rubber-mounted tank fixing bolt to free the rear of the tank. Lift and pull back on the tank to disengage the tank mounting rubbers. Place the tank in a safe place to avoid damage to the paint finish or any risk of fire.
4 Moving to the right-hand side of the machine, release the hinged flap which retains the battery. Disconnect the battery leads and drain tube and remove the battery from the machine, placing it to one side to await reassembly. Trace the alternator and CDI leads from the engine unit back to their connector block. These should be separated and all cable clips released to allow the wiring to be freed. To prevent problems during engine removal coil the leads and place them on top of the crankcase. Remove the sparking plug cap and lodge it and the HT cable clear of the engine.
5 Remove the single cross-head screw which retains the tachometer drive cable to the gearbox on the right-hand side of the cylinder head. Free the cable, lodging it clear of the engine, and refit the screw to prevent its loss. Slacken the clutch cable adjuster, displace it from its anchor bracket, and free the cable nipple from the operating arm. The decompressor cable can be released in a similar manner, or may be left in position until the right-hand cover is removed, if preferred.
6 Slacken the retaining clip which secures the carburettor to the air cleaner hose and remove the two bolts which retain the intake adaptor to the cylinder head. Twist the carburettor to free it from the cylinder head and lift it clear. Unless the carburettor requires specific attention the throttle and choke cables can be left connected and the entire assembly lodged against the frame.
7 Slacken and remove the nuts which secure the exhaust pipe flanges to the cylinder head, and displace the flanges and collet halves. Release the pillion footrest bolts to free the silencer mounting brackets to the footrest plates. Each exhaust system can now be disengaged and removed.
8 Slacken the pinch bolt which retains the gearchange linkage to the splined end of the selector shaft. Release the pivot bolt which retains the gearchange pedal to the footrest plate and remove the gearchange linkage assembly. Release the two screws which secure the gearbox sprocket cover to the casing. Lift the cover away, then remove the pressed steel guard which covers the underside of the sprocket area.
9 Slacken and remove the two sprocket mounting bolts, then turn the retainer plate until it can be slid off the splines. Release the gearbox sprocket by sliding it clear of the shaft together with the drive chain. The sprocket can now be disengaged from the chain, leaving the latter to hang around the swinging arm pivot. Should there prove to be insufficient free play in the chain to allow sprocket removal it will be necessary to slacken the rear wheel spindle and adjusters so that the necessary clearance can be obtained by moving the wheel forwards. Slacken the brake pedal pinch bolt and slide the pedal off its splines. Remove the single bolt which retains the brake pedal stop plate and lift it clear of the footrest plate.
10 Dismantle and remove the engine front mounting plate and bolts followed by the head steady assembly and the engine rear upper mounting bolt. The rear lower bolt should be slackened to allow the engine unit to pivot forward and rest on the downtube. From this point in the proceedings an assistant will be required. The engine unit can be lifted by one person if necessary, but a great deal of complicated manoeuvring can be avoided with the help of a second pair of hands. Failing this, jacks can be used to support the unit as it is released. Take the weight of the unit whilst the assistant removes the remaining mounting bolt. Lift the unit and move it forward slightly to clear the rear mountings then manoeuvre the unit clear of the frame.

4.3a Release mounting bolts and lift the dualseat away

4.3b Fuel tank is secured by single rubber-mounted bolt

4.4 Disconnect and remove the battery

4.5a Tachometer drive cable is secured by a single screw

4.5b Slacken adjuster (arrowed) and displace clutch cable from bracket

4.6 Release clips (arrowed) to free carburettor from mounting stubs

4.7 Silencers are retained by rear footrest bolts as shown

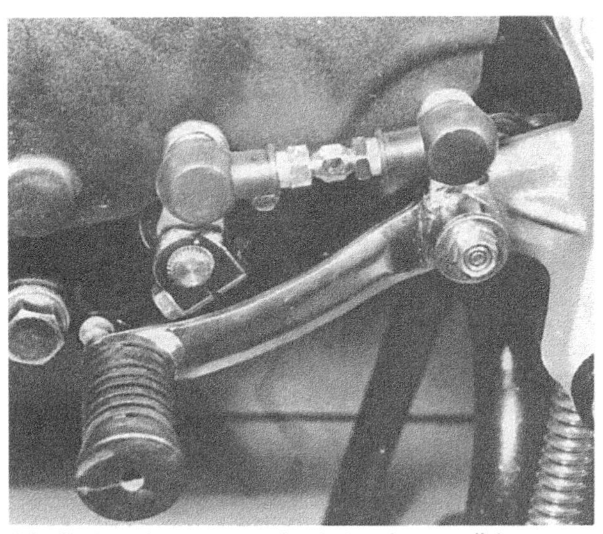

4.8a Slacken pinch bolt and pivot bolt and remove linkage

4.8b Remove the plastic sprocket cover ...

4.8c ... followed by the pressed steel guard

4.8d Gearbox sprocket can now be removed

4.10a Dismantle the engine front mounting plate, ...

4.10b ... remove cylinder head steady plates ...

4.10c ... and the upper rear mounting bolt

4.10d Lower bolt can be removed to free the engine unit

5 Dismantling the engine and gearbox: general

1 Before commencing work on the engine unit, the external surfaces must be cleaned thoroughly. A motor cycle engine has very little protection from road grit and other foreign matter, which will sooner or later find its way into the dismantled engine if this simple precaution is not observed.
2 One of the proprietary engine cleaning compounds such as 'Gunk' or 'Jizer' can be used to good effect, especially if the compound is allowed to penetrate the film of oil and grease before it is washed away. When washing down, make sure that water cannot enter the inlet or exhaust ports or the electrical system, particularly if these parts are now more exposed.
3 Never use force to remove any stubborn part, unless mention is made of this requirement in the text. There is invariably good reason why a part is difficult to remove, often because the dismantling operation has been tackled in the wrong sequence.
4 The engine unit employed in the Honda CB250 RS model is fairly straightforward in construction and poses few specific dismantling problems. The only special tool that will prove essential is an extractor for the alternator rotor. On many machines, another component from the chassis can be used as an extractor bolt, the rear wheel spindle usually proving to be ideal for this purpose. This is not true of the above model where an unusually large extractor thread is provided. If a 22 mm thread bolt is available, it can be used to good effect, but failing this the correct Honda tool, a multi-purpose extractor, will be required. Its part number is 07733-0020001.

6 Dismantling the engine/gearbox unit: removing the cylinder head cover, camshaft and cylinder head

1 As mentioned earlier in this Chapter, cylinder head removal can be carried out with the engine unit in or out of the frame. If the work is to be undertaken with the engine unit installed, it will first be necessary to remove the following items:

a) Seat and fuel tank
b) Cylinder head mounting plates
c) Carburettor
d) Exhaust pipes

Full details describing the removal of the above will be found in Section 4 of this Chapter.
2 Slacken the locknuts at the upper and lower ends of the decompressor cable. Displace the lower adjuster from its locating bracket and free the lower end of the cable from the operating lever. Repeat the procedure to release the upper end of the cable. Remove the two bolts which secure each of the valve adjuster inspection covers and lift them away. Remove the single bolt which retains the tachometer drive gearbox to the cylinder head cover and lift it away.
3 The cylinder head cover is retained by a total of 12 bolts, two of which are located inside the valve adjuster inspection holes. Note that two of the cylinder head holding bolts are located on the left-hand side of the cylinder head, recessed below the flat area surrounding the sparking plug. These should not be disturbed at this stage. Slacken the cylinder head cover bolts evenly, in a diagonal sequence to prevent warping. The cylinder head cover can now be lifted away, together with the valve rockers which are supported inside the cover.
4 Remove the larger of the two inspection plugs from the engine left-hand outer casing so that the crankshaft can be rotated by passing a socket through the inspection hole and onto the alternator rotor securing bolt. Before the camshaft and chain are removed the degree of chain wear should be assessed. The projecting part of the cam chain tensioner contains two wedges. The rearmost wedge is referred to by Honda as wedge A, whilst the second wedge, B, is nearest the camshaft sprocket. Measure the amount of wedge B which protrudes from the tensioner assembly. If this exceeds 9 mm (0.36 in) the chain will require renewal. To facilitate camshaft removal it is now necessary to relieve tensioner pressure. This is accomplished by pulling up wedge A, while pressing down wedge B, and securing it by passing a small pin or piece of wire through the 2 mm hole in wedge A. Failing this a self-grip wrench can be locked on wedge A to hold it up. It may prove helpful if an assistant can push the tensioner blade back using a screwdriver passed down the cam chain tunnel. This will remove tension from the wedges making the locking operation much easier.
5 The camshaft sprocket bolts should now be removed, turning the crankshaft to gain access to each bolt as required. Arrange the camshaft sprocket so that one of the bolt holes is uppermost with the two cutouts symmetrically disposed in relation to the gasket face. Displace the sprocket to the right until it drops clear of its locating shoulder. The cam chain can now be slid off to the left of the sprocket. Using a length of stiff wire secure the chain to prevent it from falling into the cam chain tunnel, then manoeuvre the camshaft and sprocket clear of the cylinder head.
6 Release the two cylinder head nuts. These are located on the front and rear faces of the cylinder barrel, and retain studs which project down from the cylinder head. The cylinder head is now retained by a total of four bolts. These should be slackened progressively, by about one flat at a time, in a diagonal sequence. This will ensure that the loading on the head casting remains even, avoiding any tendency to warp.
7 Slacken and remove the upper tensioner bolt and O-ring. The cylinder head can now be removed. It is quite likely that the head will be stuck to the cylinder barrel by the gasket, and if this proves to be the case it may be necessary to break the joint by tapping around the head with a soft-faced mallet. Care must be taken to avoid damage to the cooling fins. These are rather brittle and are easily chipped. For obvious reasons, do not attempt to lever the joint apart.

7 Dismantling the engine/gearbox unit: removing the cylinder barrel and piston

1 Removal of the cylinder barrel and piston can be accomplished after the cylinder head and camshaft have been removed, as described in the preceding section.
2 Remove the remaining cam chain tensioner lock bolt and plain washer and move the tensioner assembly clear of the cam chain tunnel. The cam chain guide on the exhaust side of the barrel can be lifted away. Remove the cylinder barrel retaining bolts. The cylinder barrel can now be removed by pulling it upwards. If necessary, the barrel to crankcase joint can be freed by tapping around it with a soft-faced mallet or by rocking the cylinder barrel to and fro.
3 Lift the barrel upwards by about two inches, then pack some clean rag around the crankcase mouth to catch any debris or portions of broken piston which might otherwise drop into the crankcase as the barrel is pulled clear of the piston. The barrel can now be removed completely, taking care to support the connecting rod and piston as the latter emerges from the bore.
4 Prise out one of the gudgeon pin circlips using a small electrical screwdriver or a similar tool in the slot provided. Support this piston and push the gudgeon pin out until the connecting rod is freed. The piston can now be lifted away. If the pin is a tight fit, it may be necessary to warm the piston so that the grip on the gudgeon pin is released. A rag soaked in warm water will suffice, if it is placed on the piston crown. The piston may be lifted from the connecting rod once the gudgeon pin is clear of the small-end eye.
5 If the gudgeon pin is still a tight fit after warming the piston it can be lightly tapped out of position with a hammer and soft metal drift. **Do not** use excess force and make sure the connecting rod is supported during this operation, or there is a risk of its bending.
6 When the piston is free of the connecting rod remove the gudgeon pin completely, by removing the second circlip. Place the piston, rings and gudgeon pin aside for further attention, but discard the circlips. They should never be re-used; new circlips must be obtained and fitted during rebuilding.

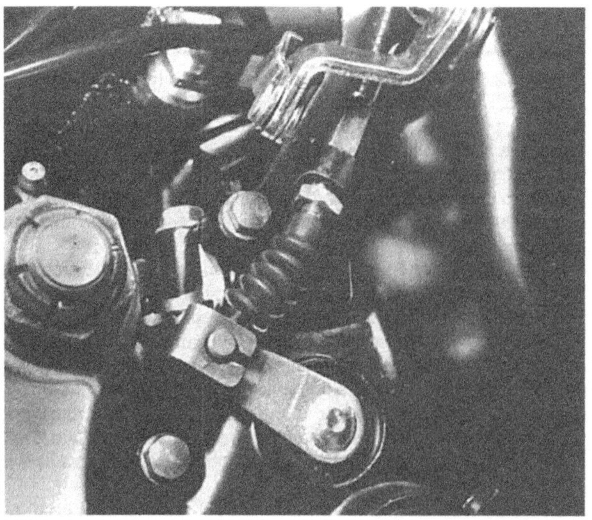

6.2a Slacken adjuster and free lower end of decompressor cable

6.2b Upper end can now be freed in a similar manner

6.2c Remove tachometer gearbox from cylinder head

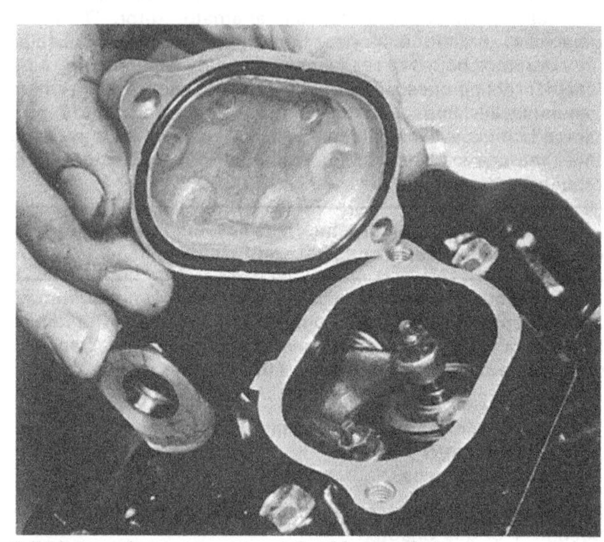

6.2d Inspection covers are retained by two bolts

6.3a Note cylinder head cover bolts 'hidden' inside casting

6.3b Cover can now be lifted clear

6.5 Release sprocket and manoeuvre camshaft clear of chain

6.6 Release the cylinder head holding nuts ...

6.7a ... and the upper tensioner lock bolt

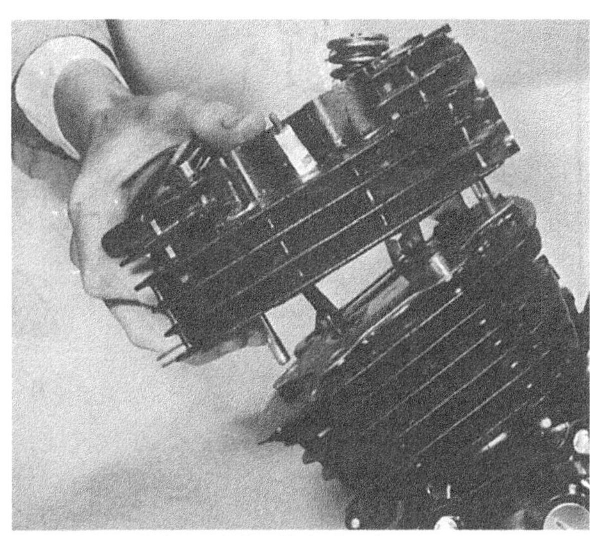
6.7b Remove bolts and lift cylinder head away

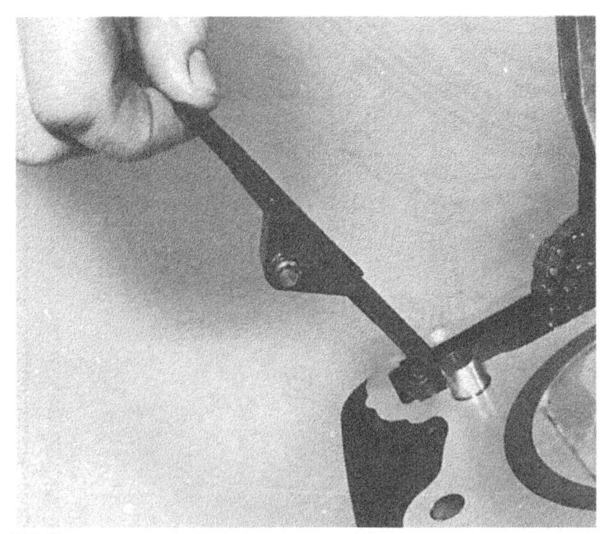
7.2 Secure cam chain and remove guide

7.4 Pad crankcase mouth, then displace circlip to free piston

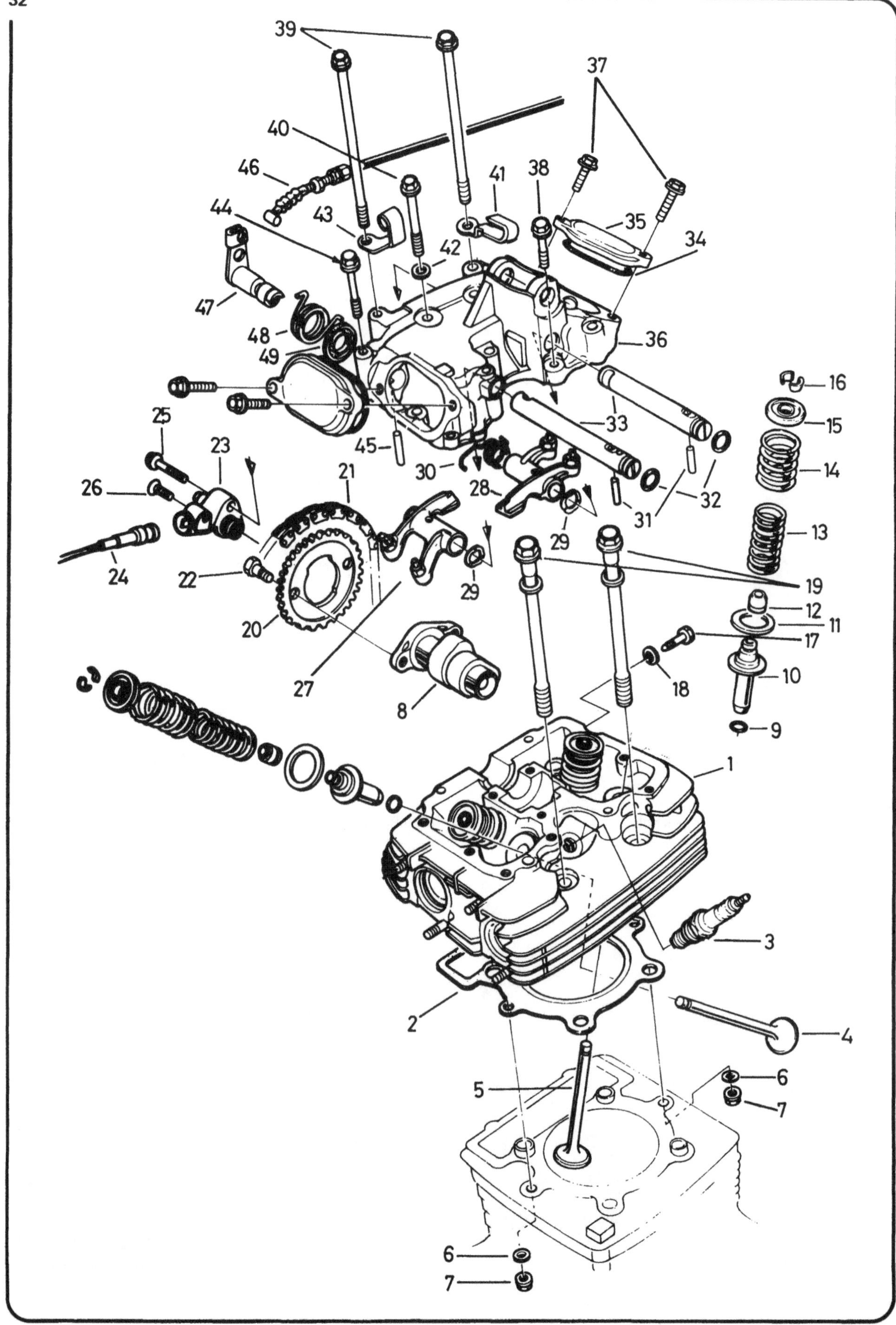

39
37
40
46
41
38
44
43
42
35
34
47
48
49
36
16
15
33
25
23
45
30
32
14
26
21
28
31
29
13
24
22
29
19
20
12
11
17
27
10
18
8
9
1
3
2
4
5
6
7
6
7

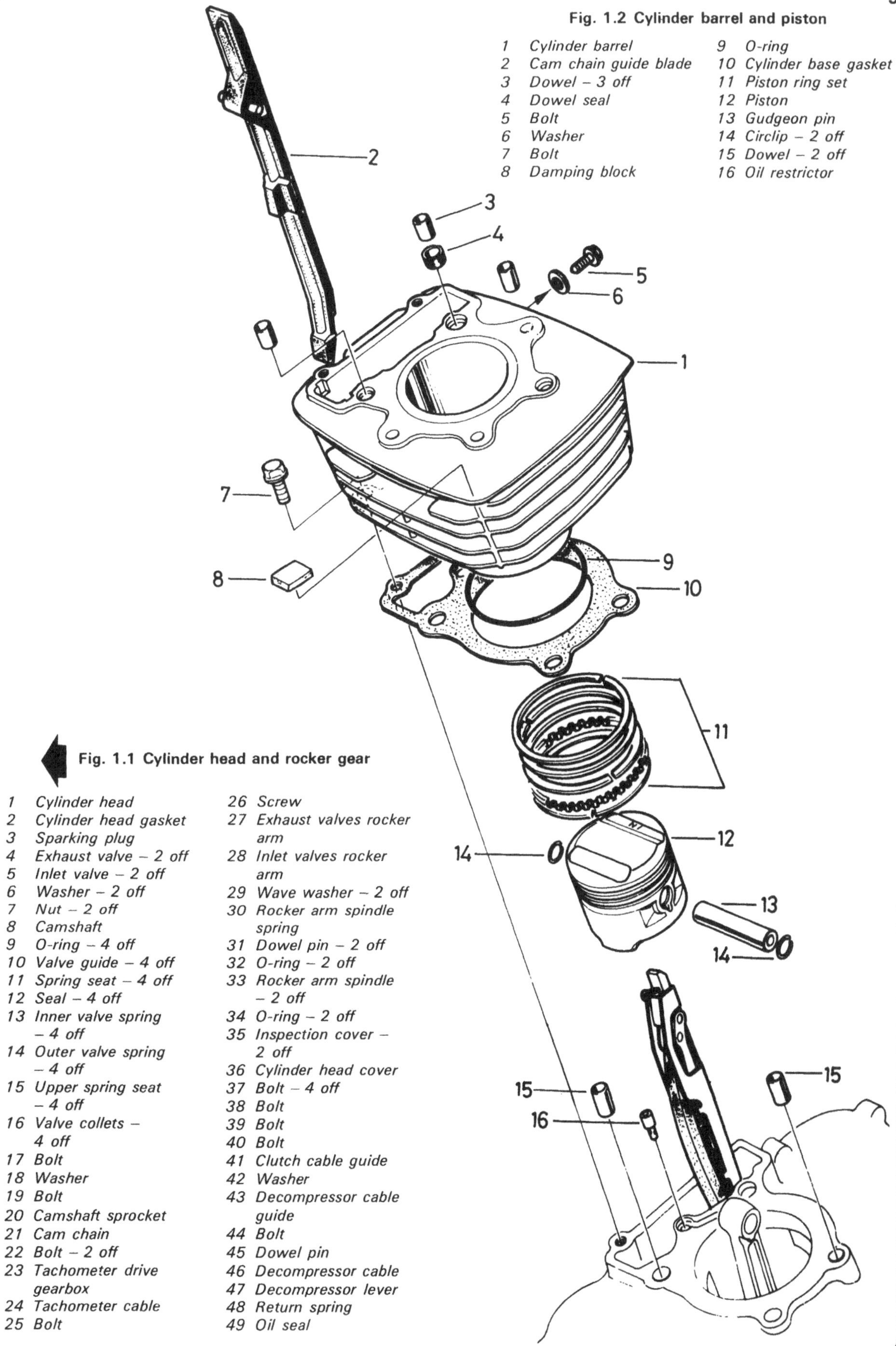

Fig. 1.2 Cylinder barrel and piston

1	*Cylinder barrel*	9	*O-ring*
2	*Cam chain guide blade*	10	*Cylinder base gasket*
3	*Dowel – 3 off*	11	*Piston ring set*
4	*Dowel seal*	12	*Piston*
5	*Bolt*	13	*Gudgeon pin*
6	*Washer*	14	*Circlip – 2 off*
7	*Bolt*	15	*Dowel – 2 off*
8	*Damping block*	16	*Oil restrictor*

Fig. 1.1 Cylinder head and rocker gear

1	*Cylinder head*	26	*Screw*
2	*Cylinder head gasket*	27	*Exhaust valves rocker arm*
3	*Sparking plug*	28	*Inlet valves rocker arm*
4	*Exhaust valve – 2 off*	29	*Wave washer – 2 off*
5	*Inlet valve – 2 off*	30	*Rocker arm spindle spring*
6	*Washer – 2 off*	31	*Dowel pin – 2 off*
7	*Nut – 2 off*	32	*O-ring – 2 off*
8	*Camshaft*	33	*Rocker arm spindle – 2 off*
9	*O-ring – 4 off*	34	*O-ring – 2 off*
10	*Valve guide – 4 off*	35	*Inspection cover – 2 off*
11	*Spring seat – 4 off*	36	*Cylinder head cover*
12	*Seal – 4 off*	37	*Bolt – 4 off*
13	*Inner valve spring – 4 off*	38	*Bolt*
14	*Outer valve spring – 4 off*	39	*Bolt*
15	*Upper spring seat – 4 off*	40	*Bolt*
16	*Valve collets – 4 off*	41	*Clutch cable guide*
17	*Bolt*	42	*Washer*
18	*Washer*	43	*Decompressor cable guide*
19	*Bolt*	44	*Bolt*
20	*Camshaft sprocket*	45	*Dowel pin*
21	*Cam chain*	46	*Decompressor cable*
22	*Bolt – 2 off*	47	*Decompressor lever*
23	*Tachometer drive gearbox*	48	*Return spring*
24	*Tachometer cable*	49	*Oil seal*
25	*Bolt*		

8 Dismantling the engine/gearbox unit: removing the right-hand outer casing and ignition pickup

1 The right-hand outer casing can be removed with the engine unit in or out of the frame. In the former instance it will be necessary to carry out the following operations prior to removal.

a) Drain the engine/transmission oil
b) Remove the kickstart lever
c) Disconnect the rear brake lamp switch spring
d) Remove the rear brake pedal and stopper plate
e) Disconnect the clutch and decompressor cables
f) Trace and disconnect the pickup leads at the connector

The various tasks mentioned above are described in Sections 4 and 6 of this Chapter.
2 Slacken and remove the hexagon-headed screws from the periphery of the casing then lift it away, taking care not to damage the kickstart shaft oil seal as it passes over the shaft splines. It is possible that a small amount of residual oil will be released as the casing is lifted clear, and some provision should be made to catch this. Take care not to lose the thrust washer which is fitted over the end of the kickstart shaft. It may drop free or stick to the casing boss and should be placed back on the shaft for safekeeping.
3 The ignition pickup coil, or pulser coil, will remain with the casing, and should be left undisturbed unless renewal is required. The ignition rotor assembly incorporates a centrifugal advance mechanism – automatic timing unit (ATU) – and is located on the crankshaft end. It can be removed after releasing the retaining nut and washer. Note that the ATU is located by a small pin in the crankshaft end. This should be removed if it is loose to preclude its loss, as should the oil feed quill and spring in the crankshaft end.

9 Dismantling the engine/gearbox unit: removing the clutch and primary drive pinion

1 The clutch and primary drive pinion can be removed with the engine in or out of the frame. Where the engine is installed a certain amount of preliminary dismantling will be required and in either case it will be necessary to remove the right-hand outer casing. Details of both will be found in Section 8 of this Chapter.
2 Remove the pushrod and thrust bearing from the centre of the clutch release plate. Slacken the four bolts which retain the release plate and lift it away. If available, the Honda clutch holding tool, part number 07923-4280000 should be fitted to the clutch centre whilst the central retaining nut is removed. Failing this, the clutch can be locked as described below.
3 Obtain two flat washers and temporarily fit them and two of the clutch bolts to compress two opposing clutch springs. The clutch will now be under tension, but the clutch centre nut will still be accessible. It will now be necessary to prevent the clutch from turning as the nut is slackened. This can be done by locking the crankshaft. If the cylinder head, barrel and piston have been removed, place a smooth round metal bar through the connecting rod small-end eye, resting the ends of the bar on suitably positioned wooden strips. The wooden strips or blocks are essential to prevent the crankcase gasket face from becoming marked.
4 An alternative to the above is to use a strap wrench fitted around the alternator rotor, having first removed the left-hand casing. The clutch can also be locked via the gearbox, by selecting top gear and restraining the gearbox sprocket. If the unit is in the frame this can be done by applying the rear brake, thus locking the entire drive train. If, however, the unit is on the workbench it will be necessary to contrive some means of holding the gearbox sprocket. The simplest method is to wrap a length of final drive chain around the sprocket and secure the ends with a self locking wrench.
5 Once the clutch assembly has been locked in position, the central retaining nut can be removed together with the Belville washer which locks it. The clutch assembly can now be pulled off its splines and placed to one side to await further examination. The primary drive pinion is located by splines on the crankshaft end and may be slid off once the automatic timing unit has been removed. The latter is retained by a nut and washer and should have been removed as described in the previous Section.

10 Dismantling the engine/gearbox unit: removing the oil pump

1 The oil pump can be removed after the clutch has been dismantled as described in the preceding Section. Start by removing the retainer plate which retains the kickstart idler gear. The plate is secured by two bolts which also pass through and retain the pump body. Lift the plate away, followed by the idler gear. The pump assembly is now clear and can be removed from the crankcase. Note that the single cross-head screw retains the pump cover to the body. It should not be removed unless it is wished to dismantle the pump.

8.3a Remove nut from crankshaft end ...

8.3b ... and slide automatic timing unit off shaft end

9.2a Remove clutch pushrod and bearing

9.2b Dismantle the clutch release plate ...

9.3a ... then fit two of the springs ...

9.3b ... with washers to lock the clutch assembly

9.5a Remove the clutch centre nut and remove clutch

9.5b Primary drive pinion can be slid off crankshaft

10.1a Oil pump assembly is retained by two bolts (arrowed)

10.1b Remove the idler gear from shaft end ...

10.1c ... to permit removal of the oil pump

11 Dismantling the engine/gearbox unit: removing the left-hand outer casing and alternator assembly

1 The alternator assembly can be removed with the engine on the workbench, or installed in the frame. In the latter instance the following operations must be carried out, using the preceding sections as reference.

a) Remove fuel tank and seat
b) Disconnect alternator output leads
c) Remove gearchange pedal
d) Remove sprocket cover
e) Drain engine oil

2 Slacken and remove the hexagon-headed screws which retain the outer casing, to the crankcase. The cover can now be lifted away complete with the alternator stator assembly which is mounted on the cover's inside face. Note that unless the stator components require renewal the assembly should be left undisturbed.

3 The rotor is mounted on the tapered end of the crankshaft. It is located by a Woodruff key and is secured by a central flanged bolt. On the machine featured in this manual the bolt was found to be extremely tight and this is likely to prove a common occurrence, particularly where the rotor is being removed for the first time since initial assembly. A secure method of preventing the rotation of the crankshaft will therefore be necessary.

4 If the engine is in the frame rotation can best be prevented by selecting top gear and locking the crankshaft through the drive train by applying the rear brake. An alternative would be to use a stout strap wrench around the inner edge of the rotor. Where the unit is being stripped prior to crankcase separation, the crankshaft can be immobilised by passing a smooth round bar through the connecting rod small-end eye and supporting the projecting ends on wooden blocks placed against the crankcase mouth. With the crankshaft restrained by one of the above methods, the securing bolt can be removed.

5 As mentioned earlier in this Chapter, it will be necessary to obtain a Honda service tool, part number 07733-0020001, to remove the flywheel rotor which will prove to be a tight fit on its taper. The only realistic alternative to the correct service tool would be a 22 mm fine pitch bolt, but this is unlikely to be found in most workshops. It should be noted that a legged puller should not be used because the rotor will almost certainly be damaged. Fit the extractor or bolt, screwing it inwards to jack the rotor clear of the crankshaft.

6 If the rotor resists removal do not continue tightening the extractor bolt down, because damage may result. Tighten the bolt fully and then strike the bolt head sharply with a hammer. This action will usually have the desired effect in breaking the fit between the tapers, but care must be exercised to avoid damage to the extractor threads when using the appropriate Honda tool.

12 Dismantling the engine/gearbox unit: separating the crankcase halves

1 Crankcase separation is necessary before attention can be given to the crankshaft, balancer shafts and drive and gearbox components. The operation can only be carried out after engine removal and preliminary dismantling as described in Sections 4 to 11 of this Chapter.

2 Working on the right-hand side of the unit, slacken and remove the balancer shaft holder lock bolt and disengage and remove the adjacent spring. Remove the tensioner anchor bolt and lift the tensioner clear of the cam chain tunnel. Displace the small oil feed pipe which runs across the centre of the tunnel to allow the cam chain to be disengaged from the crankshaft and removed. As the chain is lifted clear it should be marked clearly to indicate the outer face and the normal direction of rotation, unless a new chain is to be fitted. The incorrect fitting of a part-

worn chain can result in rapid wear of both it and the sprockets which will usually become very noisy in operation.
3 On the left-hand side of the unit, remove the two bolts which secure the balancer chain guide to the crankcase, and lift the guide away. If necessary, turn the balancer shaft holder from the right-hand end to obtain chain free play.
4 Invert the crankcase assembly on the workbench to gain access to the lower crankcase bolts. A total of eight bolts will be found and these should be slackened evenly and progressively, and then removed. Place the crankcase assembly upright once more, and remove the eight remaining crankcase bolts.
5 Once all sixteen bolts have been removed, the crankcase halves can be separated. Before proceeding with this stage it should be noted that the balancer chain runs between the upper and lower crankcase halves and will impede separation. It will be necessary to pause during separation to remove the chain from the rear balancer on the gearbox mainshaft. Some assistance in this will prove helpful.
6 If separation proves difficult, try tapping around the joint area with a soft-faced mallet. This will help to break the sometimes tenacious hold of the jointing compound. If this fails, use a hammer and a hardwood block to drive the casing halves apart, taking care to select robust areas of the casing and avoiding excessive force. Once the joint begins to separate removal is quite straightforward. Lift the upper casing half a few inches clear of the lower crankcase, then lift the gearbox mainshaft and disengage the balancer chain. The upper casing can now be placed to one side.

13 Dismantling the engine/gearbox unit: removing the front balancer, crankshaft and gear clusters

1 With the separated crankcase halves laid out on the workbench, the various components and assemblies can be removed as follows. The upper crankcase half contains the front balancer shaft and holder. Remove the single circlip which retains the balancer sprocket to the shaft end. The sprocket and drive chain can now be slid off and placed to one side. Push the shaft and holder through from the left-hand side and remove them as a unit. The oil separator plate need not be disturbed unless it requires specific attention.
2 Moving on to the lower crankcase half, remove the crankshaft assembly, noting the half-ring which locates the right-hand main bearing. Lift out the gearbox mainshaft assembly, followed by the layshaft assembly. Place these in their correct relative positions to await further dismantling or reassembly.

14 Dismantling the engine/gearbox unit: removing the gear selector mechanism

1 The external components of the gear selector mechanism can be removed with the engine unit installed, should the need arise. In the case of the selector drum and forks, however, it will be necessary to remove the engine unit and separate the crankcase halves as described in Sections 4 to 13.
2 The external components of the gearchange mechanism include the selector shaft and return spring, the selector plate and pressure spring and the detent stopper arm. All of these components operate on the protruding end of the selector drum, the tangs on the selector plate turning the drum to the next gear position when the gear change pedal is operated, and the detent stopper holding the drum in the correct position for each gear.
3 The selector shaft forms the basis of an assembly which incorporates the return spring, selector plate and its pressure spring. The assembly is removed by pulling it clear of the casing. The stopper arm is retained by a pivot bolt and incorporates a small spring which holds it against the selector drum cam. It can be removed after the bolt has been removed and spring pressure released. Care must be taken not to damage the rather frail neutral switch contact which is mounted on the end of the selector drum. If necessary, the switch contact and the selector drum cam may be removed after releasing the central retaining bolt. No further dismantling of the selector mechanism is possible until the crankcase halves have been separated.
4 Once the crankcase halves have been separated the selector fork shafts can be displaced and the selector forks removed. It is important that the forks are refitted in their original locations, so to prevent confusion during reassembly place each one on its correct shaft and in its normal position. It is also useful to degrease and mark each fork using a spirit-based felt washer.
5 The selector drum can now be removed next. It is retained at the left-hand end by a semi-circular retainer plate which is secured by two countersunk cross-head screws. These may well prove to be fairly tight and are normally coated with a thread locking compound to prevent loosening in use, so an impact driver will probably be required to avoid damage to the screw heads during removal. Once the retainer plate has been released the selector drum and bearing can be withdrawn from the casing.

11.4 Lock crankshaft to permit rotor removal

14.3a Remove stopper arm and spring ...

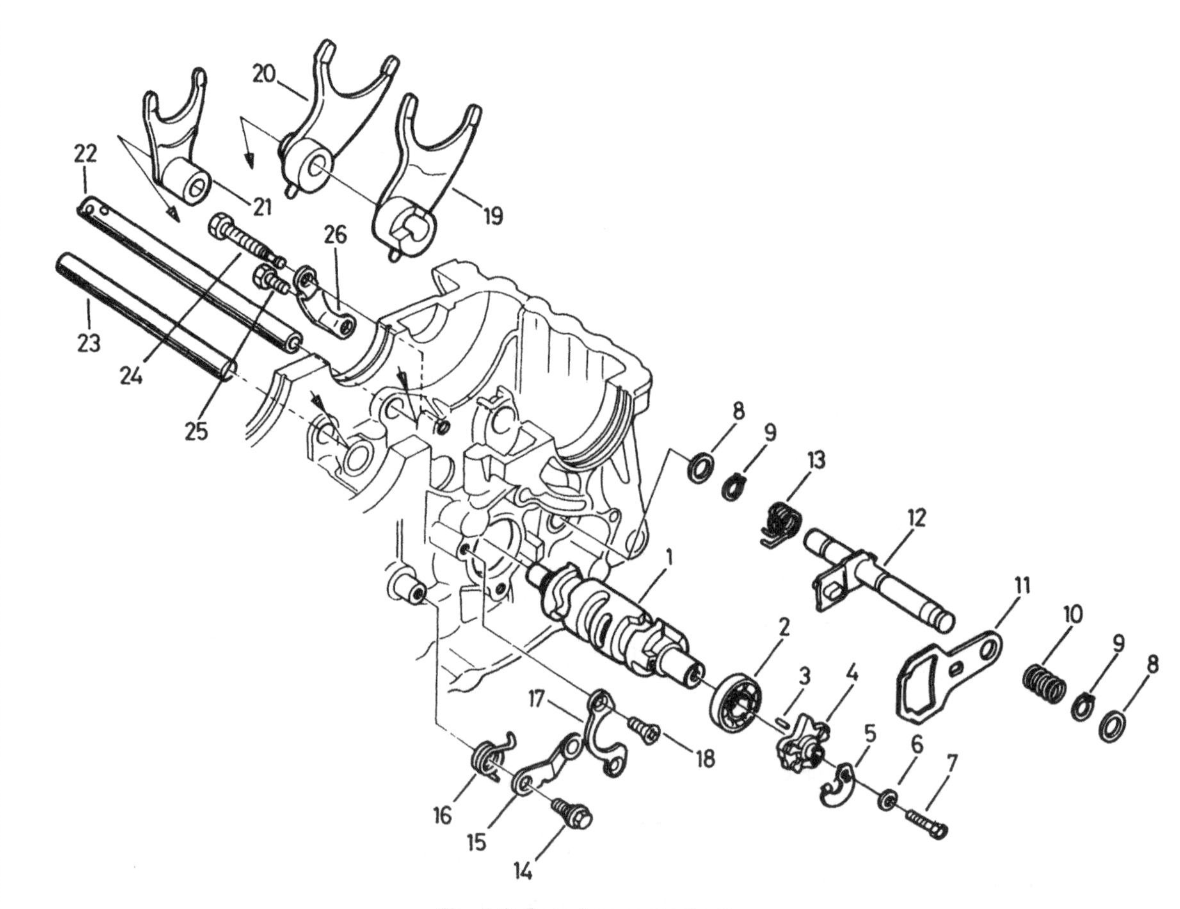

Fig. 1.3 Gearchange mechanism

1 Selector drum
2 Bearing
3 Change pin
4 Detent cam
5 Neutral switch contact
6 Washer
7 Bolt
8 Thrust washer – 2 off
9 Circlip – 2 off
10 Spring
11 Selector plate
12 Selector shaft
13 Return spring
14 Bolt
15 Stopper arm
16 Spring
17 Drum retaining plate
18 Screw – 2 off
19 Left-hand selector fork
20 Right-hand selector fork
21 Centre selector fork
22 Selector fork shaft
23 Selector fork shaft
24 Kickstart spring anchor
25 Bolt
26 Kickstart stopper plate

14.3b ... followed by switch contact and selector drum cam

14.4 Remove shafts and forks, noting their positions

14.5a Release bearing retainer from crankcase ...

14.5b ... and displace the selector drum

15 Dismantling the engine/gearbox unit: removing the kickstart shaft assembly and return spring

1 The above components can be removed only after the separation of the crankcase halves and the removal of the gearbox components as described in the preceding Sections. Start by removing the decompressor cam from the end of the kickstart shaft, together with the plain thrustwasher which precedes it. The spring and spring seat should also be slid off the shaft end.
2 Working from the inner end of the shaft, release the circlip and thrust washer from the shaft end. This will free the spring retainer which can be withdrawn from the inside of the kickstart return spring. Grasp the outer end of the spring with a strong pair of pliers, and release it from the stop. When spring tension has been released, disengage the inner tang of the spring and remove it. The shaft, pinion and ratchet can be displaced outwards and removed from the casing.

Fig. 1.4 Kickstart components

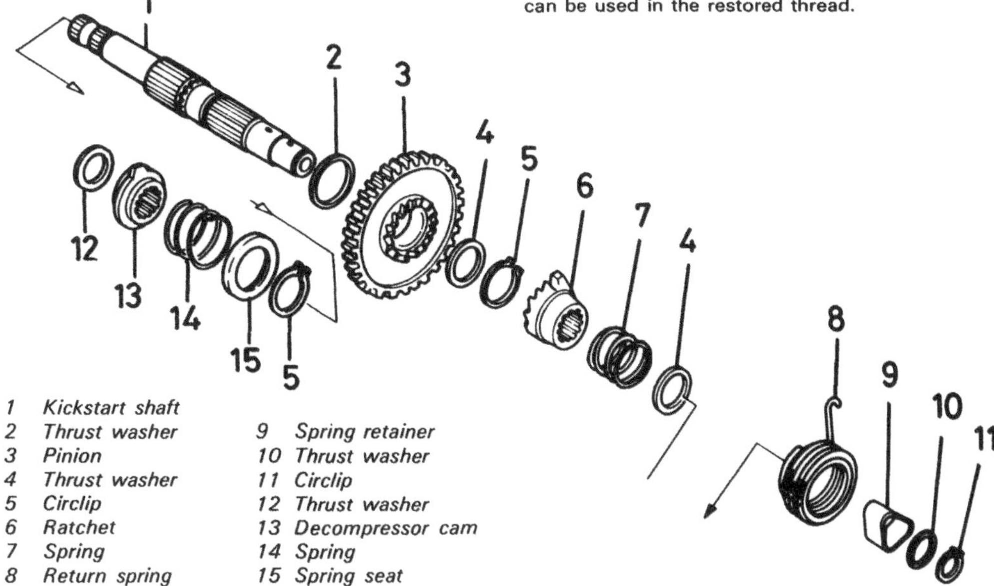

1 *Kickstart shaft*
2 *Thrust washer*
3 *Pinion*
4 *Thrust washer*
5 *Circlip*
6 *Ratchet*
7 *Spring*
8 *Return spring*
9 *Spring retainer*
10 *Thrust washer*
11 *Circlip*
12 *Thrust washer*
13 *Decompressor cam*
14 *Spring*
15 *Spring seat*

16 Examination and renovation: general

1 Before examining the parts of the dismantled engine unit for wear, it is essential that they should be cleaned thoroughly. Use a paraffin/petrol mix to remove all traces of old oil and sludge that may have accumulated within the engine.
2 Examine the crankcase castings for cracks or other signs of damage. If a crack is discovered, it will require professional repair.
3 Examine carefully each part to determine the extent of wear, if necessary checking with the tolerance figures listed in the Specifications Section of this Chapter, or accompanying the text.
4 Use a clean, lint-free rag for cleaning and drying the various components otherwise there is risk of small particles obstructing the internal oilways.
5 Should any studs or internal threads require repair, now is the appropriate time to attend to them. Where internal threads are stripped or badly worn, it is preferable to use a thread insert. The most common of these is the Helicoil type. The damaged thread is drilled oversize and then tapped to accept a diamond section wire thread insert. In most cases the original fastener can be used in the restored thread.

15.2 Release spring and guide to free kickstart assembly

17 Examination and renovation: crankcase and fittings

1 The remaining fittings on the crankcase halves should be removed prior to cleaning and examination. This applies particularly to the half-rings and dowels, which should be removed and marked by placing them in bags with suitable labels to indicate their correct location as an aid to reassembly.
2 The crankcase halves should be thoroughly degreased, using one of the proprietary water-soluble degreasing solutions such as Gunk. When clean and dry a careful examination should be made, looking for signs of cracks or other damage. Any such fault will probably require either professional repair or renewal of the crankcases as a pair. Note that any damage around the various bearing bosses will normally indicate that crankcase renewal is necessary, because a small discrepancy in these areas can result in serious mis-alignment of the shaft concerned. It is important to check crankcase condition at the earliest opportunity, because this will permit remedial action to be taken and any necessary machining or welding to be done whilst attention is turned to the remaining engine parts.
3 As mentioned previously, badly worn or damaged threads can be reclaimed by fitting a thread insert. This is a simple and inexpensive task, but one which requires the correct taps and fitting tools. It follows that the various threads should be checked and the cases taken to a local engineering works or motorcycle dealer offering this service so that repair can take place while the remaining engine parts are checked.

18 Examination and renovation: crankshaft assembly

1 Check the crankshaft assembly visually ıor damage, paying particular attention to the slot for the Woodruff key and to the threads at each end of the mainshaft. Should these have become damaged specialist help will be needed to reclaim them.
2 The connecting rod should be checked for big-end bearing play. A small amount of end float is normal, but any up and down movement will necessitate renewal.
3 Grasp the connecting rod and pull it firmly up and down. Any movement will soon become evident. Be careful that endfloat is not mistaken for wear. Should the big-end bearing be worn it will be necessary to purchase a new replacement crankshaft assembly.
4 If measuring facilities are available, set the crankshaft in V-blocks and check the big-end bearing radial clearance using a dial gauge mounted on a suitable stand. Big-end axial clearance (end float) can be checked using feeler gauges. If either clearance exceeds the service limits given in the specifications, it will be necessary to fit a new crankshaft. Honda do not supply new big-end bearings or connecting rods, so bearing replacement is not practicable.
5 Assuming that the big-end bearing is in good order, attention should be turned to the rest of the connecting rod. Visually check the rod for straightness, particularly if the engine is being rebuilt after a seizure or other catastrophe. Look also for signs of cracking. This is extremely unlikely, but worthwhile checking. Spotting a hairline crack at this stage may save the engine from an untimely end.
6 Check the fit of the gudgeon pin in the small-end eye. It should be a light sliding fit with no evidence of radial play. In the unlikely event that this condition is evident, the connecting rod, and thus the crankshaft assembly, will require renewal as no bush is fitted. It is recommended that the advice of a Honda Service Agent is sought, because he will have the necessary experience to advise on the best course of action.
7 The main bearings should be washed out with clean petrol and checked for play. Spinning the bearing on the mainshaft end will highlight any rough spots, producing obviously excessive amounts of noise once the lubricating film has been removed. Any signs of pitting or scoring of the bearing tracks or balls indicates the need for renewal. Where necessary the old bearings can be drawn off the mainshaft ends using a legged puller or bearing extractor. New bearings can be fitted using a tubular drift, but care must be taken to avoid distorting the flywheel assembly. To this end, support the adjacent flywheel half rather than risk misalignment at the crankpin. Note that the balancer drive sprocket has a punch mark which must be

18.7a Puller will be required to free main bearing from crankshaft

18.7b Alignment dot (arrowed) must coincide with keyway

aligned with the crankshaft keyway when it is refitted. When fitting the cam drive sprocket align the centre of one tooth with the dowel pin in the crankshaft. This can be aided by scribing an alignment mark on the sprocket boss.

18.7c Mark centreline of tooth and align with dowel pin

19 Examination and renovation: balancer mechanism

1 The balancer mechanism fitted to the CB250 RS model is of robust construction and is unlikely to require much attention during its normal service life which should be equal to that of the engine unit in general. Wear or damage may, however, result where the supply of lubricating oil has become limited. This could lead to excessive clearance due to wear in the needle roller bearings which support the rear balance weight or the similar bearings which carry the front balancer shaft. In extreme cases the out-of-balance forces inherent in the system could quickly destroy the bearings, causing them to break up. It will be appreciated that this is a far from desirable situation because the resulting debris would quickly destroy other engine components.

2 To check for wear during overhaul it is necessary to measure the internal diameter of the rear balance weight, and the internal and external diameters of the front balancer holder. These are as follows.

Front balancer holder	
Internal diameter	*26.007 – 26.020 mm (1.0239 – 1.0244 in)*
Wear limit	*26.05 mm (1.026 in)*
External diameter	*39.964 – 39.980 mm (1.5734 – 1.5740 in)*
Wear limit	*39.91 mm (1.571 in)*
Rear balance weight	
Internal diameter	*26.007 – 26.020 mm (1.0239 – 1.0244 in)*
Wear limit	*26.05 mm (1.026 in)*

3 The adjuster flange is retained by a large circlip to the end of the balancer holder. If it is necessary to remove the flange for any reason, make a note of its position in relation to the holder so that it can be refitted in the same place. In practice, it is unlikely that the flange will need to be disturbed.

4 Examine the balancer chain and sprocket teeth for signs of wear or damage. Neither condition is common unless the chain tension has been badly adjusted, in which case the chain will become loose and stretched and must be renewed. The sprockets must be renewed if the teeth are obviously hooked or chipped. On no account should worn or damaged sprockets be reused in view of the risk of rapid chain wear or breakage that might result. The chain should be renewed together with the sprockets to ensure long and reliable service.

20 Examination and renovation: cylinder barrel

1 The usual indications of a badly worn cylinder barrel and piston are excessive oil consumption and piston slap, a metallic rattle that occurs when there is little or no load on the engine. If the top of the bore of the cylinder barrel is examined carefully, it will be found that there is a ridge on the thrust side, the depth of which will vary according to the amount of wear that has taken place. This marks the limit of travel of the uppermost piston ring.

2 If possible, an internal micrometer should be used to obtain an accurate measurement of the amount of wear which has taken place in the bore. Measurements should be made at various points within the bore, and the measurement across the most worn part of the bore compared with that of an unknown portion, e.g. the lowest part of the bore. The difference between the two readings should not exceed 0.10 mm (0.004 in). If it is found to exceed this figure, it will be necessary to have the cylinder rebored and a new oversize piston fitted.

3 If, as is likely, an internal micrometer is not available, a rough check of the amount of bore wear can be made as follows. Insert the bare piston into the bore in its normal position ie, with the arrow facing the front of the cylinder. Using feeler gauges, measure the amount of clearance between the piston and bore about $\frac{1}{2}$ in from the top of the cylinder liner, and at the front of the cylinder. If this measurement exceeds the allowable clearance as stated above, it indicates that attention is required. If desired, the barrel and piston can be taken to a Honda Service Agent for verification.

4 Check the surface of the cylinder bore for score marks or any other damage that may have resulted from an earlier engine seizure or displacement of the gudgeon pin. A rebore will be necessary to remove any deep indentations, irrespective of the amount of bore wear, otherwise a compression leak will occur.

5 Check the external cooling fins are not clogged with oil or road dirt; otherwise the engine will overheat.

20.4 Check bore for wear and scoring

21 Examination and renovation: piston and piston rings

1 If a rebore is necessary, the existing piston and rings can be disregarded because they will be replaced with their oversize equivalents as a matter of course.

2 Remove all traces of carbon from the piston crown, using a soft scraper to ensure the surface is not marked. Finish off by polishing the crown with metal polish so that carbon does not adhere so easily in the future. Never use emery cloth.

3 Piston wear usually occurs at the skirt or lower end of the piston and takes the form of vertical streaks or score marks on the thrust side. There may also be some variation in the thickness of the skirt.

4 The piston ring grooves may also become enlarged in use, allowing the piston rings to have greater side float. If the clearance exceeds the piston ring to groove clearance wear limit figures given in the Specifications Section, the piston will require renewal, complete with rings. It should be noted that it is unusual for this type of wear to be found in an otherwise unworn engine.

5 The piston rings will tend to lose their elasticity over a period of time, the eventual result being that they will allow combustion pressure to escape into the crankcase, thus causing a noticeable drop in power. It is recommended that the free end gap of the rings be measured, and compared with the figures given in the Specifications Section. The rings should be renewed as a set if the free end gap is lower than that given under 'Wear limit'.

6 Piston ring wear is measured by removing the rings from the piston and inserting them in the cylinder bore using the crown of the piston to locate them approximately $1\frac{1}{2}$ inches from the top of the bore. Make sure they rest square with the bore. Measure the ring end gap using feeler gauges, and compare this reading with that given in the Specifications Section. Note that it is assumed that the cylinder is not in need of a rebore, as a worn bore would produce an inaccurate reading.

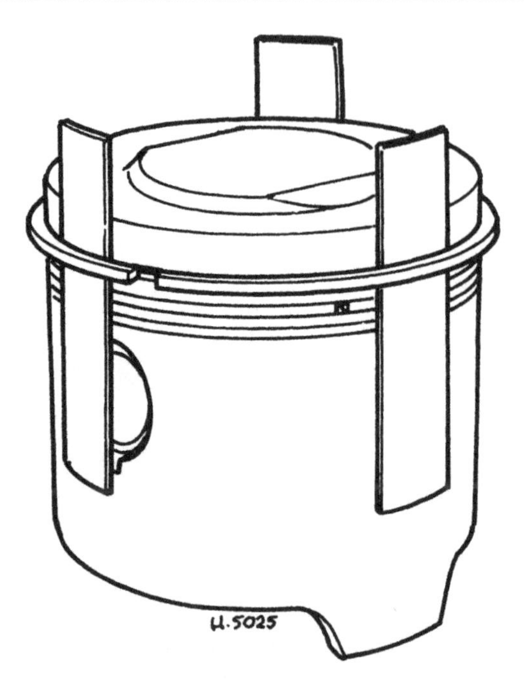

Fig. 1.5 Method of freeing gummed piston rings

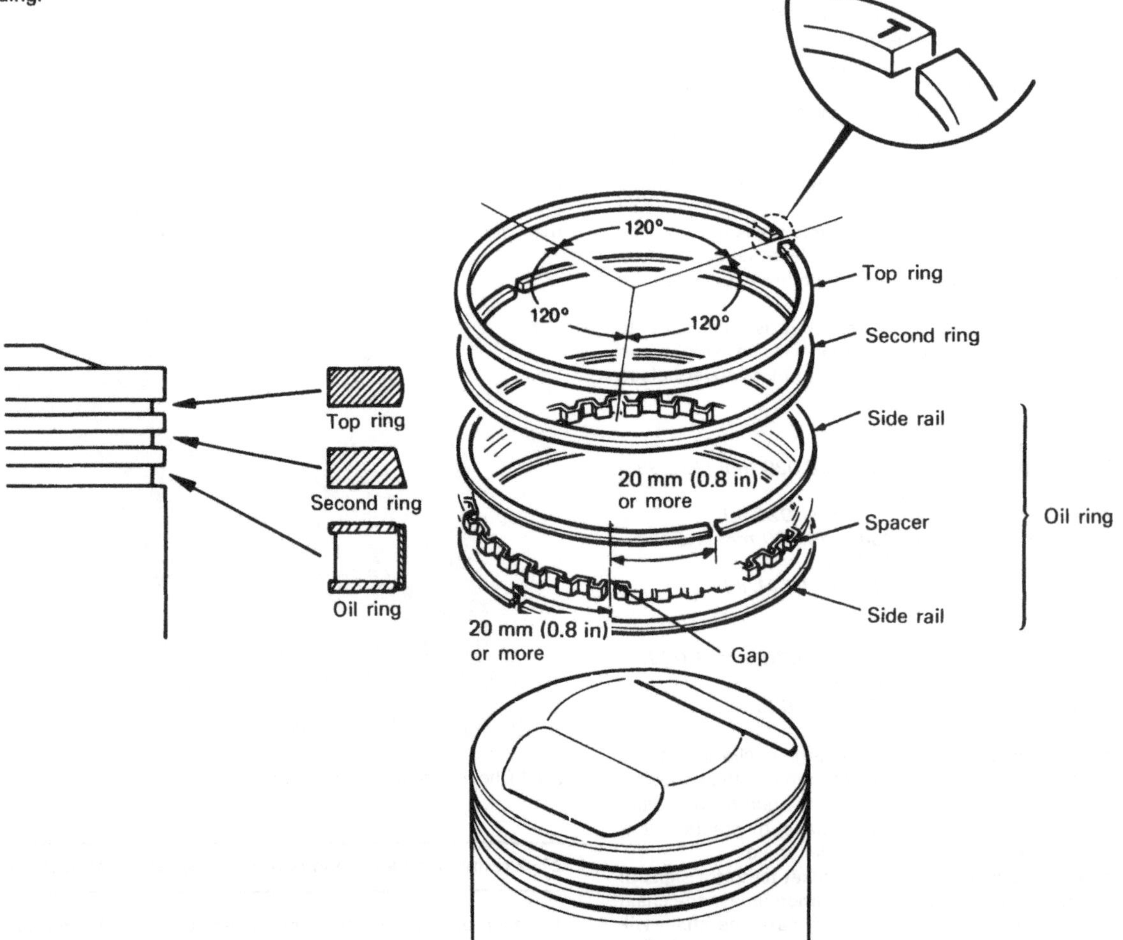

Fig. 1.6 Piston ring end gap positions and profiles

21.4 Piston should be free from scores or scorch marks

22 Examination and renovation: cylinder head and valves

1 It is best to remove all carbon deposits from the combustion chambers before removing the valves for inspection and grinding in. Use a blunt end chisel or scraper so that the surfaces are not damaged. Finish off with a metal polish to achieve a smooth, shining surface. If a mirror finish is required a high speed felt mop and polishing soap may be used. A chuck attached to a flexible drive will facilitate the polishing operation.
2 A valve spring compression tool must be used to compress each set of valves in turn, thereby allowing the split collets to be removed from the valve cap and the valve springs and caps to be freed. Keep each set of parts separate so that they can be replaced in the correct location. This is best done by placing each valve, springs and spring seat in a marked plastic bag or similar to indicate its correct position. It is essential that the pairs of inlet valves and exhaust valves do not become interchanged.
3 Before attention is turned to the valves and valve seats it is important to check for wear between the valve stems and guides. The appropriate nominal and service limit figures can be found in the Specifications Section of this Chapter. Measure the valve stem at the point of greatest wear and then measure again at right-angles to the first measurement. If the valve stem is below the service limit it must be renewed. The valve stem/guide clearance can be measured with the use of a dial gauge and a new valve. Place the new valve into the guide and measure the amount of shake with the dial gauge tip resting against the top of the stem. If the amount of wear is greater than the wear limit, the guide must be renewed.
4 To remove an old valve guide, place the cylinder head in an oven and heat it to about 100°C, taking care to ensure warpage does not occur through overheating or uneven heating. The old guide can now be tapped out from the cylinder side. The correct drift should be shouldered with the smaller diameter the same size as the valve stem and the larger diameter slightly smaller than the OD of the valve guide. If a suitable drift is not available a plain brass drift may be utilised with great care. To aid removal of the valve guide and to help prevent broaching of the cylinder head material, all carbon deposits on the portion of the valve guide projecting into the port should be cleaned off prior to guide removal. Each valve guide is fitted with an oil seal to ensure perfect sealing. The oil seal must be replaced with new components. New guides should be fitted with the head at the same heat as for removal. Note that after fitting new guides the valve seats must be recut using a 45° cutter or stone to ensure that each seat is correctly aligned with the guide axis.
5 Valve grinding is a simple task. Commence by smearing a trace of fine valve grinding compound (carborundum paste) on the valve seat and apply a suction tool to the head of the valve. Oil the valve stem and insert the valve in the guide so that the two surfaces to be ground in make contact with one another. With a semi-rotary motion grind in the valve head to the seat, using a backward and forward action. Lift the valve occasionally so that the grinding compound is distributed evenly. Repeat the application until an unbroken ring of light grey matt finish is obtained on both valve and seat. This denotes the grinding operation is now complete. Before passing to the next valve, make sure that all traces of the valve grinding compound have been removed from both the valve and its seat and that none has entered the valve guide. If this precaution is not observed rapid wear will take place due to the highly abrasive nature of the carborundum paste.
6 When deep pits are encountered, it will be necessary to use a valve refacing machine and a valve seat cutter, set to an angle of 45°. Never resort to excessive grinding because this will only pocket the valves in the head and lead to reduced engine efficiency. If after cutting the seal it is found that the seat width exceeds 2.0 mm, the width should be restored to the correct range of 1.2 – 1.4 mm (0.05 – 0.06 in) using first a 32°, and then a 60° cutter or stone. Because of the expense of such equipment and the expertise required for its satisfactory use it is recommended that seat restoration be placed in the hands of an expert. If there is any doubt about the condition of a valve, fit a new one.
7 Examine the condition of the valve collets and the groove on the valve stem in which they seat. If there is any sign of damage, new parts should be fitted. Check that the valve spring collar is not cracked. If the collets work loose or the collar splits whilst the engine is running, a valve could drop into the cylinder and cause extensive damage.
8 The condition of the valve springs can be assessed by measuring their free lengths and comparing the readings with those specified. If the engine is being overhauled after many miles of use, it is usually worthwhile renewing the valve springs as a matter of course.
9 Reassemble the valve and valve springs by reversing the dismantling procedure. Fit new oil seals to each valve guide and oil both the valve stem and the valve guide, prior to reassembly. Take special care to ensure the valve guide oil seal is not damaged when the valve is inserted. As a final check after assembly, give the end of each valve stem a sharp tap with a hammer, to make sure the split collets have located correctly.
10 Check the cylinder head for straightness, especially if it has shown a tendency to leak oil at the cylinder head joint. If there is any evidence of warpage, provided it is not too great, the cylinder head must be either machined flat or a new head fitted. Most cases of cylinder head warpage can be traced to unequal tensioning of the cylinder head nuts and bolts by tightening them in incorrect sequence.
11 Mild cases of cylinder head warpage can be cured by lapping the cylinder head on a sheet of abrasive paper placed on a surface plate or a sheet of plate glass. It will be necessary to remove the two projecting studs prior to this operation. Move the head with an oscillatory movement over the abrasive paper, removing no more than the minimum amount of cylinder head material required to correct the warpage. Once the mating face is flat, finish off using a very fine abrasive paper to produce a smooth matt grey finish.

23 Examination and renovation: rocker arms and spindles

1 The Honda CB250 RS model employs two forked rocker arms to operate the four valves. The arms pivot on harden steel spindles which are located in the cylinder head cover. Each spindle is retained by a small dowel pin which passes through its left-hand end via the cylinder head cover casting. It is theoretically possible to remove the pins by grasping them with a pair of pliers. It has been found in practice that the pins are usually a very tight fit in the rocker shafts, and can prove very difficult to move. The pins are of lightly polished hardened steel

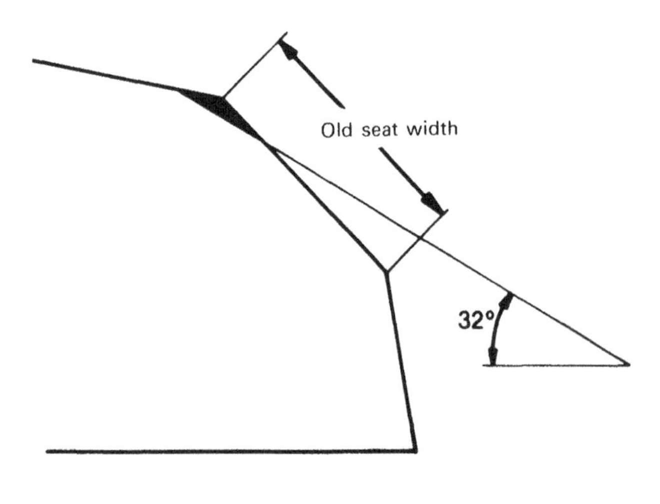

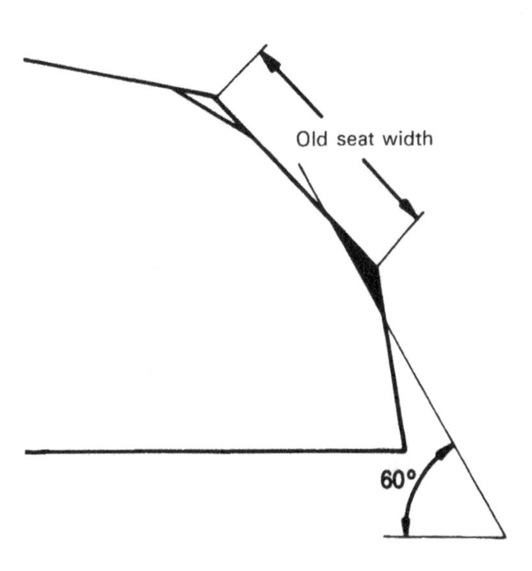

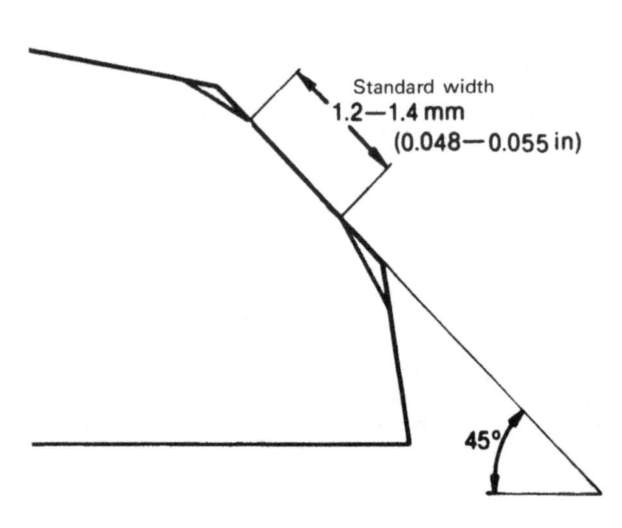

Fig. 1.7 Valve seat re-cutting angles

and are difficult to grasp firmly enough to permit removal. On occasions it may prove helpful to heat the cover by immersing it in boiling water to assist in the removal operation. An alternative approach, provided at least 4 mm of the pin protrudes above the casing, is to grind a slot in the side of the pin to a depth of about 2 mm. This will prov:de a purchase point for the tip of a screwdriver which may be used as a lever. A high-speed rotary grinder with a knife-edge will be required because of poor access to the pins.

2 After the dowel pins have been removed, the spindles can be displaced and the rockers and thrust washers released. It may prove useful to tap the cylinder head cover to jar the spindles clear, using a soft-faced mallet to avoid damage to the soft alloy. Remove the rockers and spring washers and place them back on their respective shafts to avoid any part becoming interchanged.

3 Check the rocker arms for undue wear on their spindles and renew any that show excessive play. Examine each rocker arm where it bears on the cam and the opposite end which bears on the valve stem head. Arms that are badly hammered or worn should be renewed. Slight wear marks may be stoned out with an oil carborundum stone, but remember that if too much metal is removed it will not only weaken the component but may make correct valve clearance adjustment difficult. Further, the depth of hardening may be exceeded and, therefore, subsequent wear will be rapid.

22.2a Compress valve springs and remove split collets

22.2b Upper spring seat can now be removed ...

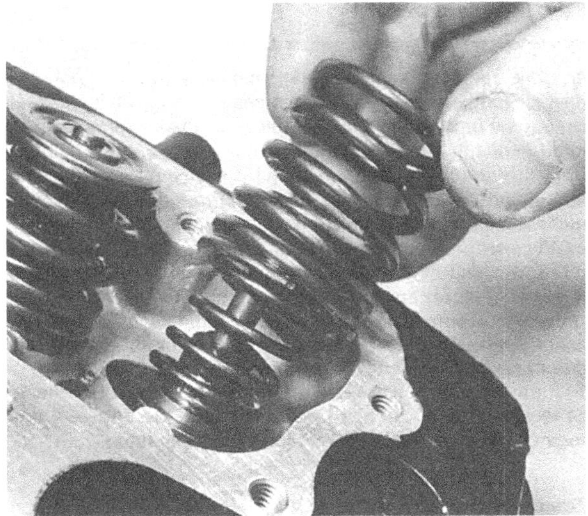
22.2c ... followed by inner and outer valve springs ...

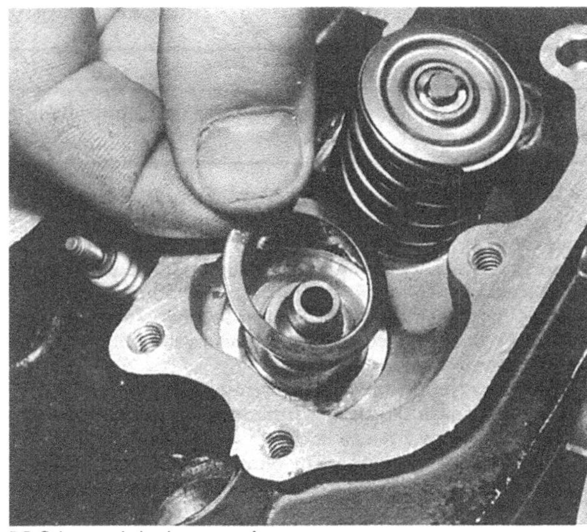
22.2d ... and the lower spring seat

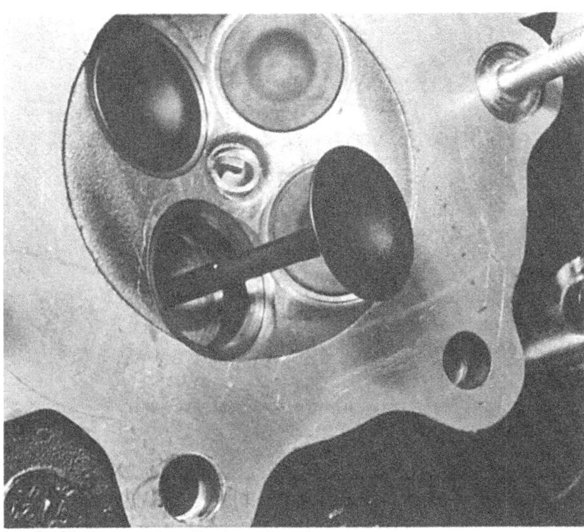
22.2e Valve can be displaced and removed

22.9 Note valve guide oil seals

23.1 Rocker shafts are secured by dowel pins (arrowed)

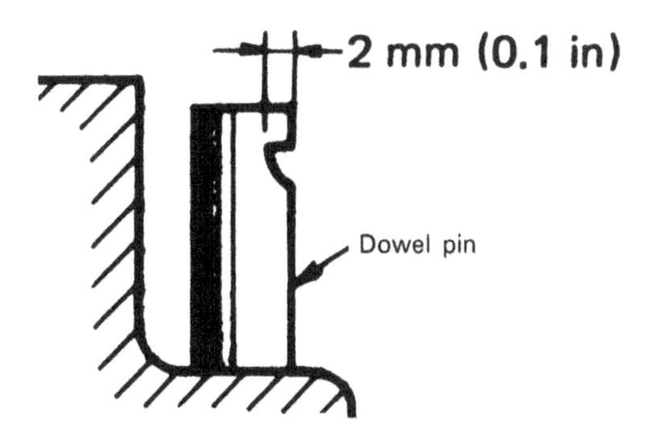

Fig. 1.8 Removal of rocker spindle locating dowels

24 Examination and renovation: camshaft and bearing surfaces

1 The camshaft should be examined visually for wear, which will probably be most evident on the ramps of each cam and where the cam contour changes sharply. Also check the bearing surfaces for obvious wear and scoring. Cam lift can be checked by measuring the height of the cam from the bottom of the base circle to the top of the lobe. If the measurement is less than the service limit the opening of that particular valve will be reduced resulting in poor performance.
2 Measure the diameter of each bearing journal with a micrometer or vernier gauge. If the diameter is less than the service limit, renew the camshaft,
3 The camshaft bears directly on the cylinder head material, there being no separate bearings. Check the bearing surfaces for wear and scoring. The clearance between the camshaft bearing journals and the aluminium bearing surfaces may be checked by measuring the journals and bearing surfaces, using an internal and external micrometer and calculating the amount of clearance. If the clearance is greater than given for the service limit the recommended course is to replace the camshaft. If bad scuffing is evident on the camshaft bearing surfaces of the cylinder head, due to a lubrication failure, the only remedy is to renew the cylinder head, and the camshaft if it transpires that it has been damaged also.
4 In the case of older machines that are well out of their warranty period, it may be worth considering a conversion to needle roller camshaft bearings as an alternative to the renewal of the cylinder head, cover and camshaft. This will offer a saving over the cost of renewing the worn parts and has the advantage of being easily and inexpensively repaired should the bearings wear in the future. A number of engineering companies will undertake this type of work, and often advertise in motorcycle magazines. Note that this type of conversion is not undertaken by or recommended by Honda.

25 Examination and renovation: cam chain, sprockets and tensioner

1 After high mileages have been covered it may become apparent that the cam chain tensioner is unable to compensate for wear and stretch in the cam chain, leading to noisy operation. The tensioner assembly is automatic in operation, the blade moving under spring tension to take up any slack appearing in the cam chain. A wedge arrangement prevents the tensioner adjustment from backing off. This arrangement can normally be expected to last between overhauls, but if the cam chain operation becomes noisy it indicates that the chain or tensioner is in need of renewal.
2 As already mentioned, the wedge nearest the cam sprocket (wedge B) provides a quick means of checking chain wear prior to dismantling. If it has risen to 9 mm (0.36 in) or more above the tensioner body it can be assumed that the chain is in need of renewal. If the chain has stretched or the tensioner blade worn to the point where the tensioner assembly cannot compensate, either or both parts will require renewal. If the engine has been dismantled for a general overhaul it is usually worthwhile renewing the chain as a precautionary measure, together with the tensioner blade if it appears worn. Note that where a part-worn chain is to be re-used it must be refitted so that it runs in the same direction as it did prior to removal. If the chain direction is reversed, accelerated wear and noisy operation may result.
3 Wear in the tensioner blade or the cam chain guide will be self evident, and if either component is deeply grooved it should be renewed. In extreme cases, a combination of a badly worn tensioner blade and guide will prevent adjustment of a serviceable chain. On no account should the pad material be allowed to wear through to the backing metal.
4 The camshaft mounted cam chain sprocket is bolted in position and, in consequence, is easy to remove if the teeth become worn, chipped or broken. The lower sprocket is an interference fit on the crankshaft and, therefore, must be pulled from position if renewal is required. Because of its positioning and shape difficulty may be encountered when trying to gain purchase on the sprocket using a legged puller. If this proves to be the case the main bearing should be displaced, taking with it the sprocket. Refer to Section 18, paragraph 7. When refitting the lower sprocket ensure that the centre line of any one tooth is in exact alignment with the dowel pin in the crankshaft. If this is not done, correct valve timing will not be possible. Scribing a datum line on the shaft will help in achieving accuracy.
5 If the sprocket(s) are renewed, the chain should be renewed at the same time. It is bad practice to run old and new parts together since the rate of wear will be accelerated.

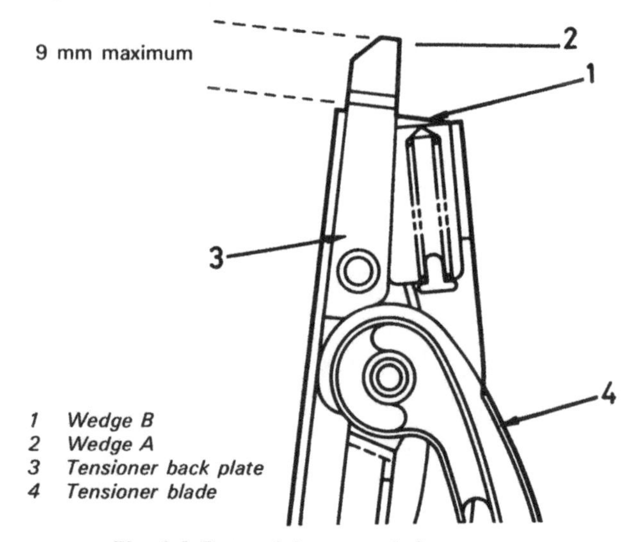

Fig. 1.9 Determining cam chain wear

26 Examination and renovation: clutch assembly

1 After an extended period of service the clutch linings will wear and promote clutch slip. The limit of wear measured across each inserted plate and the standard measurement is as follows:

	Standard	Service limit
Clutch plate thickness	*2.62 – 2.78 mm (0.102 – 0.109 in)*	*2.3 mm (0.09 in)*

When the overall width reaches the limit, the inserted plates must be renewed, preferably as a complete set.
2 The plain plates should not show any excess heating (blueing). Check the warpage of each plate using plate glass or a surface plate and a feeler gauge. The maximum allowable warpage is 0.30 mm (0.012 in).
3 The clutch springs will lose tension after a period of use, and should be renewed as a precaution if clutch slip has been evident and the friction plates are within limits. The free length of the clutch springs give a good indication of condition, and this should be checked and compared with the figures given in the Specifications Section.
4 Examine the clutch assembly for burrs or indentations on the edges of the protruding tongues of the inserted plates and/or slots worn in the edges of the outer drum with which they engage. Similar wear can occur between the inner tongues of the plain clutch plates and the slots in the clutch inner drum. Wear of this nature will cause clutch drag and slow disengagement during gear changes, since the parts will become trapped

and will not free fully when the clutch is withdrawn. A small amount of wear can be corrected by dressing with a fine file; more extensive wear will necessitate renewal of the worn parts.
5 The clutch release mechanism takes the form of a spindle running in the right-hand outer casing, the shaped end of which bears on the clutch release pushrod when the handlebar lever is operated. The mechanism is of robust construction and requires no attention during normal maintenance or overhauls.

27 Examination and renovation: gearbox components

1 Examine each of the gear pinions to ensure that there are no chipped or broken teeth and that the dogs on the end of the pinions are not rounded. Gear pinions with any of these defects must be renewed; there is no satisfactory method of reclaiming them.
2 The gearbox bearings must be free from play and show no signs of roughness when they are rotated. The bearings should first be washed in petrol and then dried. Also check for pitting on the roller tracks.
3 It is advisable to renew the gearbox oil seals irrespective of their condition. Should a re-used seal fail at a later date, a considerable amount of work is involved to gain access to renew it.
4 Check the gear selector rods for straightness by rolling them on a sheet of plate glass. A bent rod will cause difficulty in selecting gears and will make the gear change particularly heavy.
5 The selector forks should be examined closely, to ensure that they are not bent or badly worn. Under normal conditions, the gear selector mechanism is unlikely to wear quickly, unless the gearbox oil level has been allowed to become low.
6 The tracks in the selector drum, with which the selector forks engage, should not show any undue signs of wear unless neglect has led to under lubrication of the gearbox. Check the tension of the gearchange pawl, gearchange arm and drum stopper arm springs. Weakness in the springs will lead to imprecise gear selection. Check the condition of the gear stopper roller and the pins in the change drum end with which it engages. It is unlikely that wear will take place here except after considerable mileage.
7 Check the condition of the kickstart components. If slipping has been encountered a worn ratchet and pawl will invariably be traced as the cause. Any other damage or wear to the components will be self-evident. If either the ratchet or pawl is found to be faulty, both components must be replaced as a pair. Examine the kickstart return spring which should be renewed if there is any doubt about its condition.
8 If it proves necessary to dismantle the gearbox shafts for further examination or to renew worn or damaged parts, reference should be made to the accompanying line drawing and the photographic sequence which shows the correct assembly sequence (see Section 29). When rebuilding the shafts it is advisable to use new thrust washers and circlips throughout, and these should be obtained when purchasing the new seals and gaskets required for reassembly.

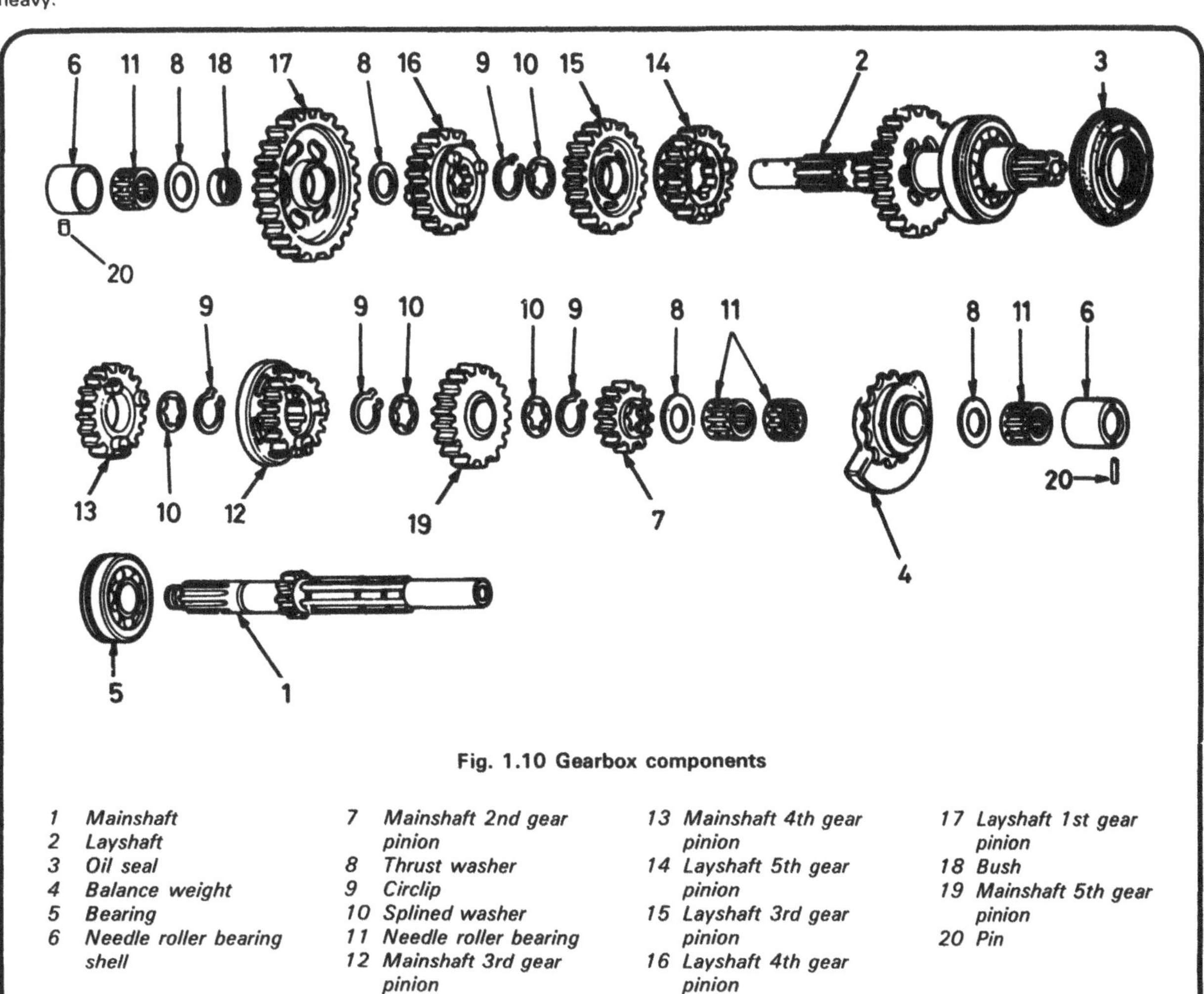

Fig. 1.10 Gearbox components

1 Mainshaft
2 Layshaft
3 Oil seal
4 Balance weight
5 Bearing
6 Needle roller bearing shell
7 Mainshaft 2nd gear pinion
8 Thrust washer
9 Circlip
10 Splined washer
11 Needle roller bearing
12 Mainshaft 3rd gear pinion
13 Mainshaft 4th gear pinion
14 Layshaft 5th gear pinion
15 Layshaft 3rd gear pinion
16 Layshaft 4th gear pinion
17 Layshaft 1st gear pinion
18 Bush
19 Mainshaft 5th gear pinion
20 Pin

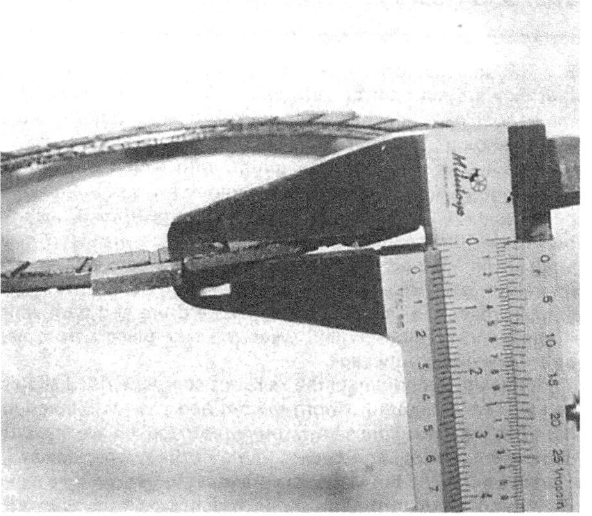
26.1 Measure thickness of clutch friction plates

26.4 Clutch rattle can develop if shock absorber springs become weak or broken

26.5 Release mechanism rarely requires attention

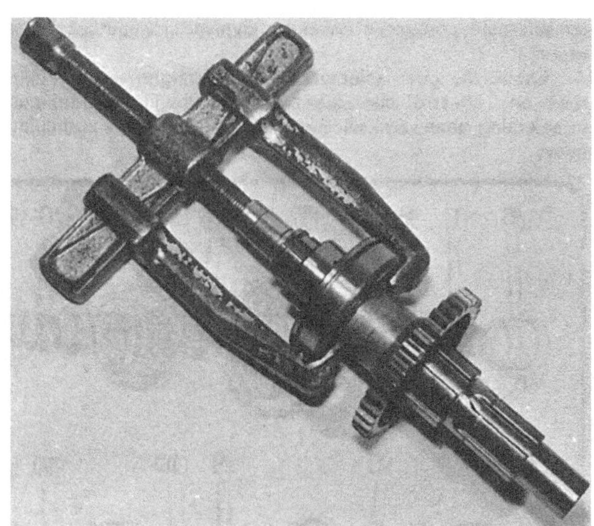
27.2 Puller is needed to remove gearbox bearings

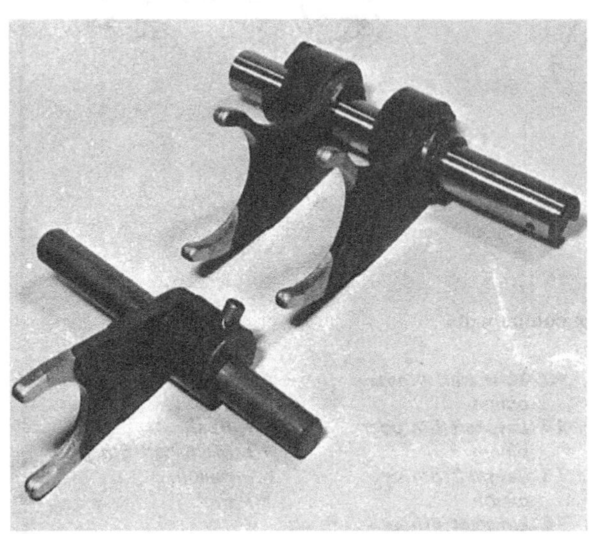
27.5 Check forks for wear. Note position of each fork

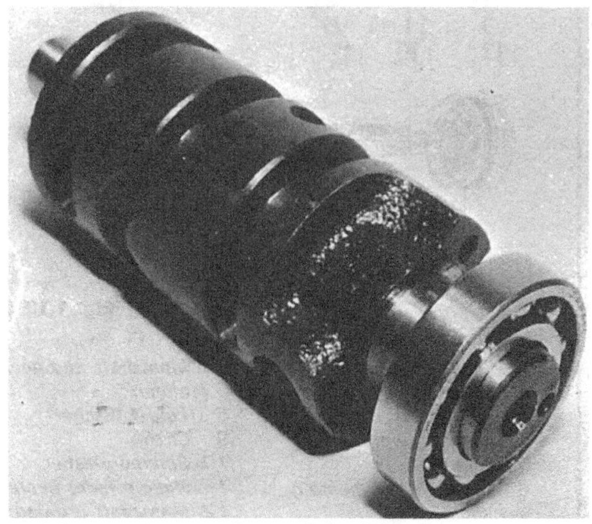
27.6 Selector drum bearing is a light sliding fit

28 Engine reassembly: general

1 Before reassembly of the engine/gear unit is commenced, the various component parts should be cleaned thoroughly and placed on a sheet of clean paper, close to the working area.
2 Make sure all traces of old gaskets have been removed and that the mating surfaces are clean and undamaged. One of the best ways to remove old gasket cement is to apply a rag soaked in methylated spirit. This acts as a solvent and will ensure that the cement is removed without resort to scraping and the consequent risk of damage.
3 Gather together all of the necessary tools and have available an oil can fitted with clean engine oil. Make sure all new gaskets and oil seals are to hand, also all replacement parts required. Nothing is more frustrating than having to stop in the middle of a reassembly sequence because a vital gasket or replacement has been overlooked.
4 Make sure that the reassembly area is clean and that there is adequate working space. Refer to the torque and clearance settings wherever they are given. Many of the smaller bolts are easily sheared if over-tightened. Always use the correct size screwdriver bit for the crosshead screws and never an ordinary screwdriver or punch.

29 Engine reassembly: rebuilding the gearbox clusters

1 If the gearbox components have been dismantled for examination and renewal, it is essential that they are rebuilt in the correct order to ensure proper operation of the gearbox. Use the accompanying photographic sequence and line drawing as aids to the identification of components and their correct relative positions.

Layshaft (output shaft) assembly
2 Take the layshaft complete with its left-hand bearing and 2nd gear pinion, and fit the 5th gear pinion with its selector fork groove facing the right-hand end of the shaft.
3 Slide the 3rd gear pinion into position, followed by the splined thrust washer. Retain them with a circlip.
4 The 4th gear pinion is fitted next, with its selector groove facing inwards (towards the left), followed by a plain thrust washer.
5 Fit the large 1st gear pinion, ensuring that it engages the dogs of the preceding gear. Do not omit the separate gear pinion bush.
6 Place the right-hand layshaft bearing (needle roller type) over the shaft end, preceded by the thrust washer.
7 Fit a new oil seal to the left-hand end of the shaft.

Mainshaft (input shaft) assembly
8 Where necessary, fit a new bearing to the right-hand (clutch) end of the mainshaft, ensuring that it butts squarely against the integral 1st gear pinion. Note that the locating groove must face outwards.
9 Place the 4th gear pinion in position, ensuring that the dogs face towards the left-hand end of the shaft. Place a splined thrust washer against the 4th gear pinion and secure it with a circlip.
10 The 3rd gear pinion should be fitted next, with its selector groove facing inwards, or towards the right-hand end of the shaft.
11 This is followed by a circlip and a thrust washer, and then the 5th gear pinion. Fit another thrustwasher and circlip to retain it.
12 Slide the small 2nd gear pinion into place followed by the large plain thrustwasher which locates it.
13 Assembly is completed by fitting the bearings and the balance weight. A wide needle roller bearing is fitted first. This is followed by a narrow needle roller bearing. Fit the balancer weight with the sprocket facing outwards, then finish the assembly by fitting the outer needle roller bearing preceded by the remaining large thrust washer.
14 Moving to the opposite end of the shaft, fit the flanged spacer with the widest diameter inwards, then place the clutch bearing in position.

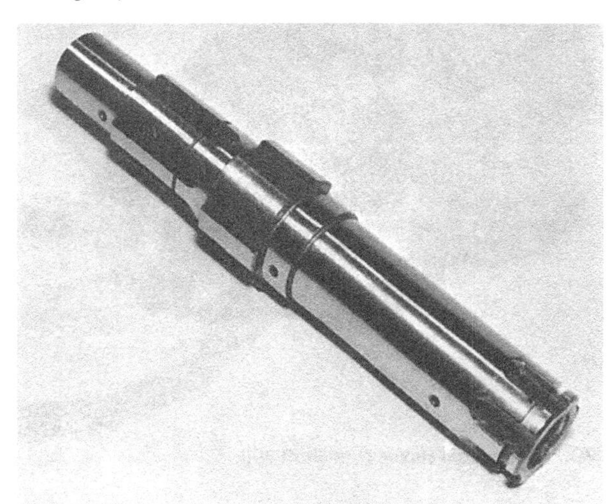
29.2a The bare layshaft (output shaft)

29.2b Fit the 2nd gear pinion ...

29.2c ... followed by small and large spacers

29.2d Bearing should be pressed onto shaft, groove outwards

29.2e A length of steel tubing can be used to fit bearing

29.2f Slide steel sleeve over shaft end

29.2g Fit the 5th gear pinion with groove outwards as shown

29.3a Recessed face of 3rd gear pinion faces inwards

29.3b Fit splined washer and retain with circlip

29.4a Fit the 4th gear pinion with selector groove inwards

29.4b Place thrustwasher against shoulder ...

29.5 ... and fit the 1st gear pinion and bush

29.6 Assemble thrustwasher and needle roller bearing

29.8 RH mainshaft bearing is press fit against integral 1st gear

29.9a Position 4th gear pinion with dogs arranged as shown

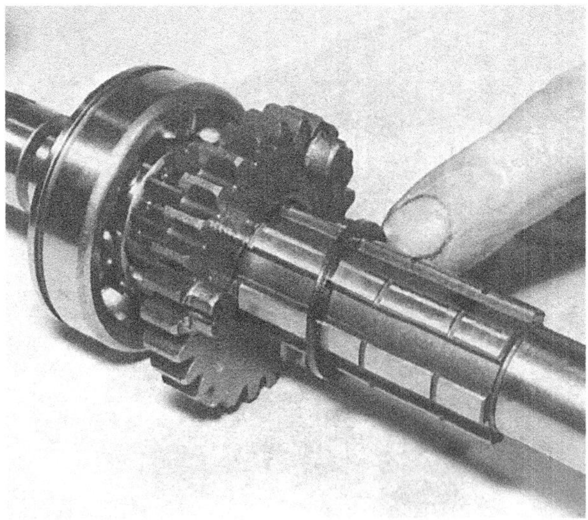
29.9b Fit a splined washer ...

29.9c ... and secure with circlip

29.10 3rd gear pinion is positioned as shown

29.11a Fit circlip into adjacent groove ...

29.11b ... followed by splined washer ...

29.11c ... and 5th gear pinion

29.11d Secure with splined washer and circlip

29.12a 2nd gear can now be slid into position ...

29.12b ... followed by large plain thrust washer

29.13a Fit bearings and balance weight as shown ...

29.13b ... followed by thrustwasher ...

29.13c ... and needle roller bearing

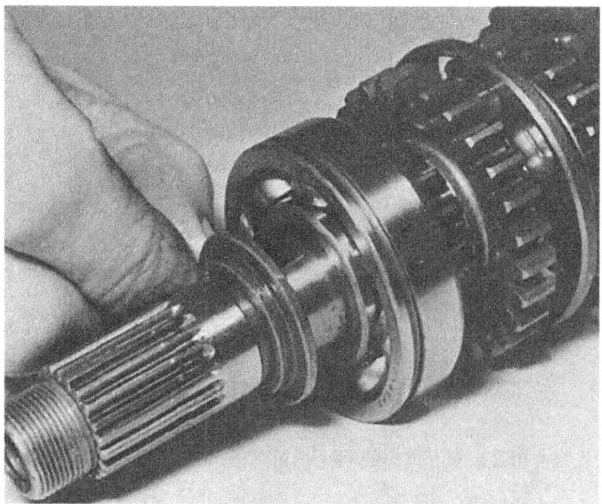
29.14a Note the position of flanged spacer against bearing

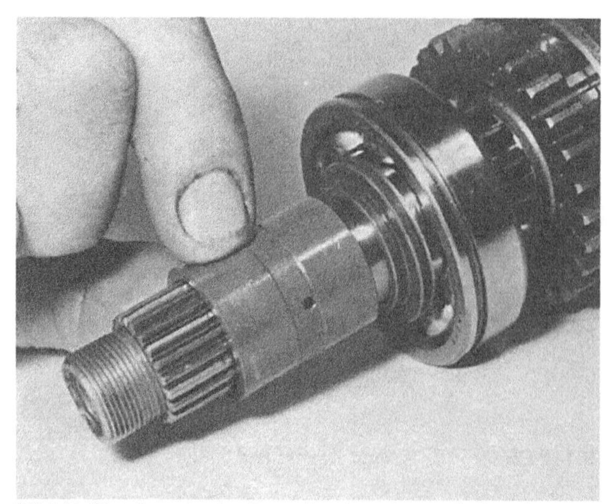
29.14b Clutch centre bush can be placed over shaft end

30 Engine reassembly: refitting the half-rings, dowels and kickstart mechanism

1 Place the lower crankcase half on the workbench and fit the locating half-rings to the grooves provided in the left-hand layshaft boss and the right-hand mainshaft and crankshaft bosses. Place the locating dowels in the appropriate holes at the front and rear of the jointing surface.
2 Refit the kickstart stopper plate to the outer face of the lower crankcase half and refit the spring anchor pin. The pin is secured by a nut on the inside of the casing. The anchor pin and the retaining bolt should be tightened to the following torque figures:

Anchor pin	*2.2 – 3.0 kgf m (16 – 22 lbf ft)*
Anchor pin lock nut	*1.6 – 2.0 kgf m (12 – 15 lbf ft)*
Retaining bolt	*0.8 – 1.2 kgf m (6 – 9 lbf ft)*

3 Fit the larger circlip, thrustwasher, starter pinion, thrust washer and the smaller circlip to the kickstart shaft, securing the pinion to the shaft. Place the starter ratchet over the shaft splines, ensuring that the alignment marks on the shaft and the ratchet are correctly positioned. If this is not set accurately, correct kickstart operation will not be possible. Fit the ratchet spring and thrust washer.
4 Fit the shaft assembly into the casing, ensuring that the spring hole at the inner end faces upwards and that the oil hole is visible in its casing recess. Before proceeding further, fill the oil hole with clean engine oil. Place the return spring over the shaft and engage its inner tang in the cross-drilling in the shaft. Once engaged, fit the retaining collar between the spring and shaft.
5 Using a stout pair of pliers, turn the free end of the spring until it can be hooked over the anchor pin. Fit the plain washer against the retaining collar and secure it with a new circlip, ensuring that its chamfered edge faces outwards. It is advisable to renew this circlip as a precaution, in view of the considerable dismantling work that would be required should it become displaced in use. Check that it seats correctly.
6 Moving to the external part of the kickstart shaft assembly, place the spring seat and spring over the protruding end of the shaft. Fit the decompressor cam, ensuring that the punch marks on the cam and shaft are correctly aligned.

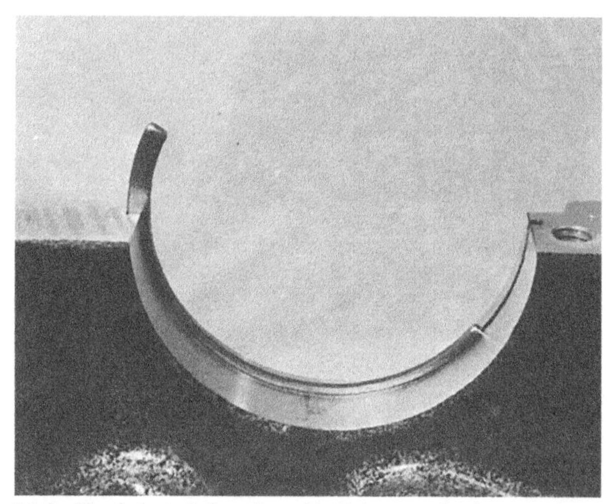
30.1a Check that all dowels and half rings are in place

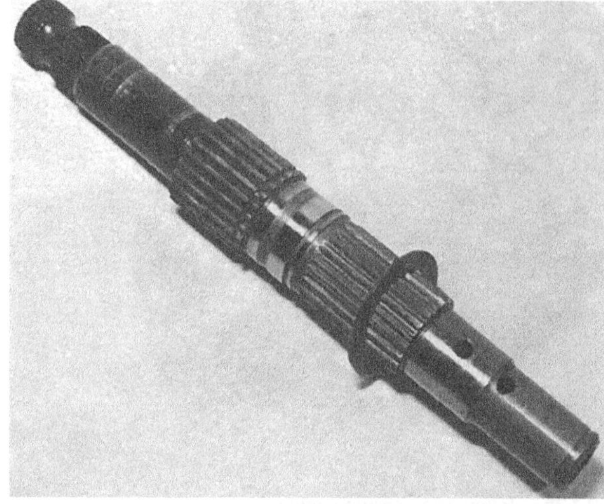
30.3a Fit thrust washer and circlip to outer end of shaft

30.3b Slide kickstart pinion into position ...

30.3c ... followed by thrust washer ...

30.3d ... and retain with circlip

30.3e Fit ratchet noting alignment dots

30.3f Place ratchet spring and plain washer over shaft

30.4 Slide assembly into casing and locate spring and guide

30.5 Tension spring and fit plain washer and circlip

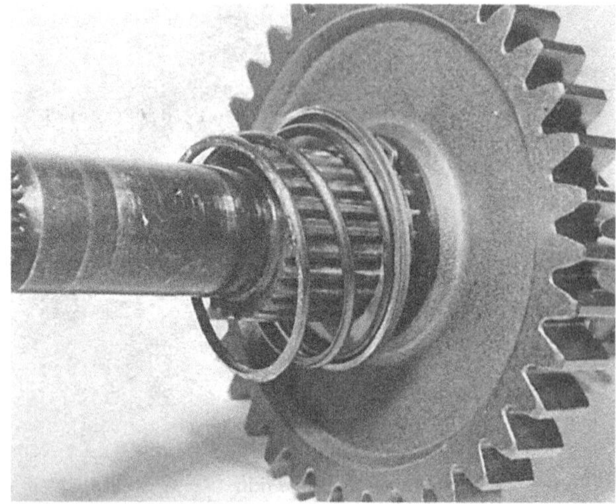

30.6a Fit spring seat and spring to outer end of shaft

30.6b Slide cam into place, aligning the punch marks (arrowed)

31 Engine reassembly: refitting the gear selector mechanism

1 Fit and lubricate the selector drum bearing, and lubricate the plain end of the selector drum. The drum can now be slid into the casing ensuring that the bearing locates correctly. Fit the bearing retainer plate, using a thread locking compound on the two countersunk screws which secure it.

2 Slide each of the selector fork shafts into the casing, positioning the forks on the shafts as they enter the casing bores. Make sure that the forks are arranged as shown in the accompanying photograph and line drawing.

3 Place the stopper plate over the projecting end of the selector drum, noting the small dowel pin which locates the stopper plate in relation to the drum. Assemble the spacer, neutral switch contact and retaining bolt and washer. Note that it is possible to fit the neutral switch contact in two positions, and care should be taken to ensure that it is fitted correctly. The outer face of the stopper plate has four raised dogs with which the selector claws engage. The remaining lobe of the plate is plain, and the neutral contact end should coincide with the first dog clockwise from the plain lobe.

4 Turn the selector drum until the plain lobe of the stopper plate is almost vertical, then assemble the detent stopper arm and spring, securing it with its pivot bolt. The roller on the end of the stopper arm should engage the small neutral recess in the plate. Check that the selector shaft is correctly assembled, then install it in the casing, taking care not to damage the neutral contact during installation.

5 Temporarily refit the gearchange pedal and check that the mechanism selects each gear correctly. It will be necessary to hold each of the forks in engagement with the selector drum tracks, and these should be lubricated with engine oil to ensure free movement. When the selection has been checked, set the drum back to the neutral position.

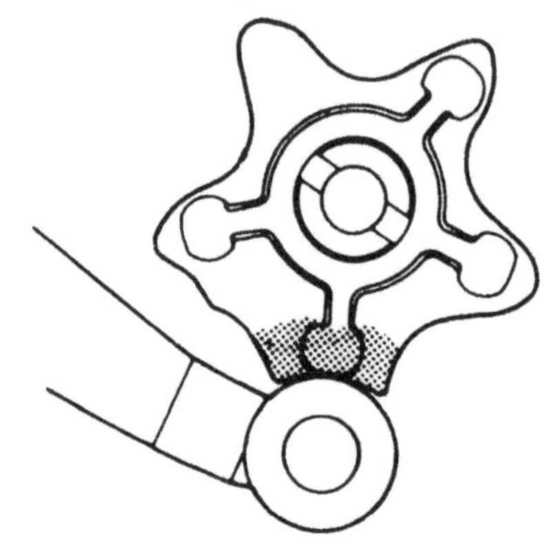

Fig. 1.11 Gearchange drum in neutral gear position

31.2 Fit the selector forks and support shafts

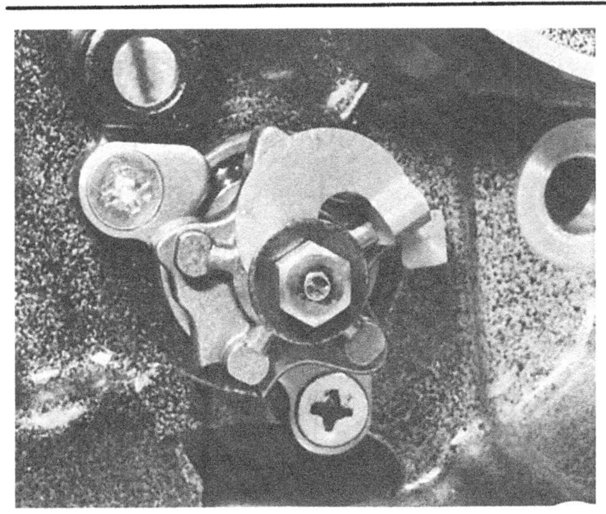
31.3 Note position of neutral switch contact

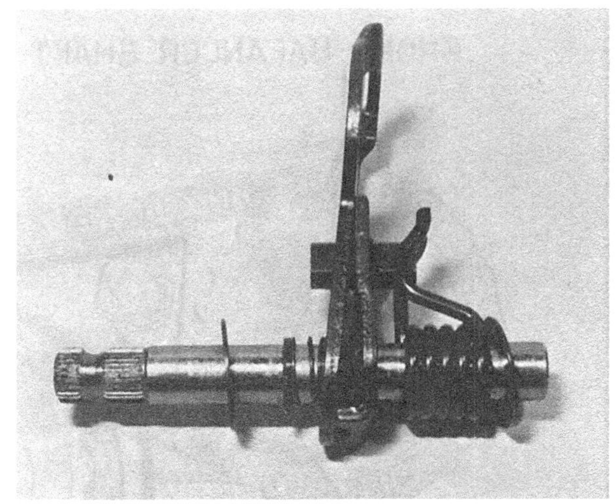
31.4a Check that selector shaft is assembled as shown ...

31.4b ... then install, taking care not to damage contact

32 Engine reassembly: refitting the gearbox clusters and crankshaft

1 Place the gearbox layshaft (output shaft) cluster in position, ensuring that the locating lip on the oil seal engages with its groove and that the half-ring locates in the bearing groove. The small needle-roller bearing on the right-hand end of the shaft should engage with the small locating dowel at the bottom of the casing recess, and to assist in location a pair of scribed lines on the end of the outer race should align with the gasket face.
2 Place the balancer chain around the rear balance weight sprocket, ensuring that the white-painted timing links face outwards. If the marks are indistinct or missing, lay the chain on the workbench and stretch it into an elongated loop. Mark a link at the left-hand end of the lower run, then, working from left to right, mark the 9th and 27th links in a similar fashion. Note that for timing purposes, the 1st link should coincide with the front balancer mark, the 9th link with the crankshaft and the 27th link with the rear balancer mark. The chain should therefore be arranged so that the 27th link coincides with the timing dot on the rear balancer. The gearbox mainshaft (input shaft) should now be fitted, bearing in mind the above.
3 Lower the crankshaft assembly into position in the crankcase, making sure that the bearing seat squarely and that the half-ring locates in the right-hand main bearing.

32.2a Install the gearbox mainshaft assembly ...

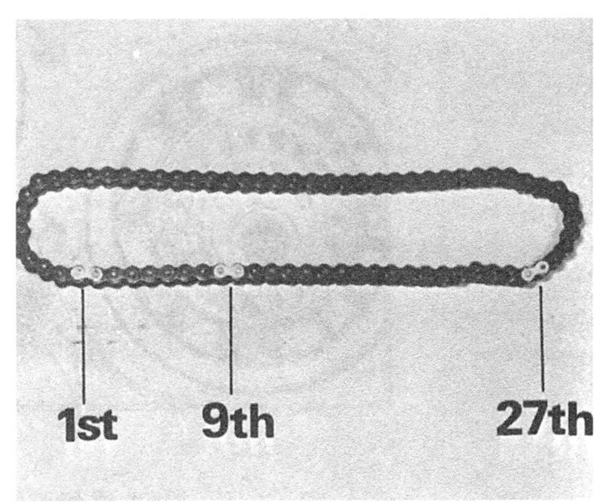

32.2b ... aligning 27th link with timing dot

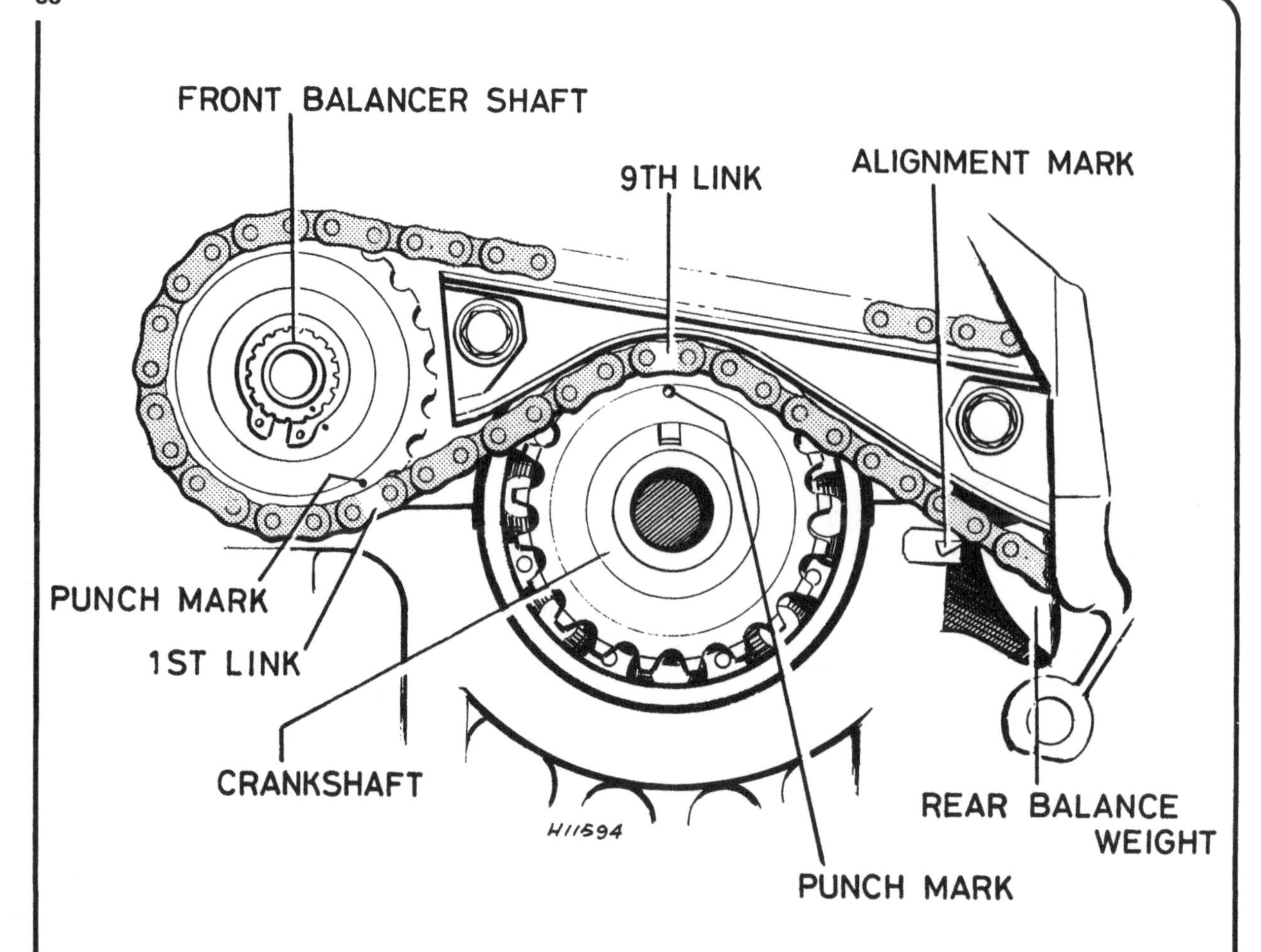

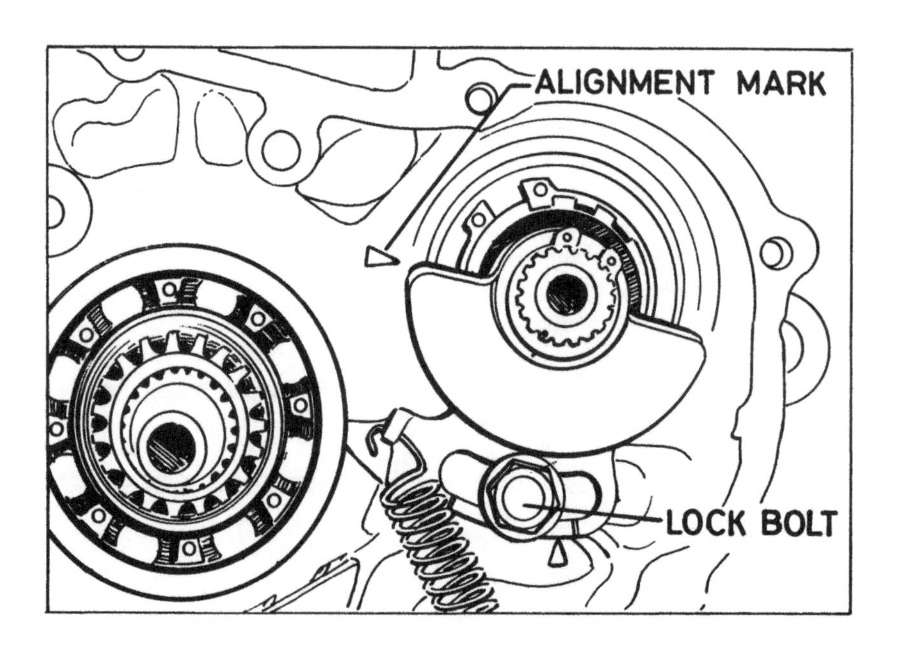

Fig. 1.12 Balancer chain timing marks

32.2c Lower layshaft assembly into casing ...

32.3 ... and position the crankshaft assembly

33 Engine reassembly: refitting the front balancer assembly and setting the balancer timing

1 There are two points to note prior to reassembly. Firstly, it is necessary to assemble the balancers and drive chain and to check the timing as the crankcase halves are joined. This is because the front balancer is housed in the upper half, whilst the rear balancer is mounted on the end of the gearbox mainshaft, and is thus contained in the lower half. To overcome this problem it is easiest to assemble the crankcase dry, then after the balancer timing has been set the cases can be parted just sufficiently for the jointing compound to be applied. Secondly, the chain must be in position on the front balancer sprocket **before** it is secured to the shaft, and similarly it must engage the rear balancer sprocket before the mainshaft is fully home in the casing. It was found in practice that the chain could be fitted around either sprocket at the outset, and then fitted around the remaining sprocket as reassembly progressed. Each method has its own drawbacks and advantages but either may be used. The method shown in the accompanying photographs involved the fitting of the chain around the rear balancer first.

2 Start by assembling the front balancer shaft components. This can easily go wrong if care is not taken, and it is recommended that the following sequence and the accompanying illustrations are studied carefully before work commences. Fit a circlip to each of the two grooves on the balancer shaft. Where possible, new circlips should be used to preclude any risk of displacement in service. The clips are machine-stamped and will have one flat face and one with rounded edges. The former (flat) must always be positioned to take any thrust loading, which means in this instance that the flat faces should be positioned towards the centre of the shaft. Now place the special tanged thrustwashers against the circlips, ensuring that the circlip ends fit between two adjacent tangs. Note that this point must be rechecked at intervals during reassembly, because the washers tend to be easily displaced.

3 Place a needle roller bearing against each thrustwasher and lubricate both with engine oil. It will be noted that the shaft ends are of different lengths when measured to the thrustwashers. The balance weight is fitted to the longer end. Place the large plain washer against the needle roller bearing, then slide the weight over the splined end of the shaft ensuring that the punch marks on each component are in alignment. Secure the weight with a new circlip. Again, make sure that the clip is fitted correctly with the rounded edge inwards.

4 If it has been removed for any reason, refit the adjusting flange on the end of the balancer holder, positioning it in the centre of the range of adjustment slots and securing it with its circlip. The shaft and balance weight can now be slid into place from the adjusting flange end. Take care that the tanged thrustwasher at the inner end of each bearing does not become displaced.

5 Slide the balancer holder into the crankcase upper half, noting that the weight should be on the **right-hand** side of the casing. With the casing inverted on the workbench it is easy to get this wrong, so double check during installation.

6 Check that all the lower casing components are correctly positioned and that the marked 27th link of the chain is aligned with the index mark on the rear balancer sprocket. Turn the crankshaft to TDC and place the chain across the top so that the 9th link aligns the dot on the crankshaft sprocket.

7 Place the upper crankcase half loosely on top of the assembled lower half. Place the front balancer sprocket in the loop of chain extending from the rear sprocket, ensuring that the timing mark aligns with the paint-marked 1st link. Fit the thrust washer to the balancer shaft end, then align the marks on the sprocket and shaft and slide the sprocket into place. The sprocket can be secured by its circlip.

8 The crankcase halves should now be assembled 'dry' to check the balancer timing. Set the crankshaft at TDC by pulling the connecting rod fully upwards. Carefully lower the upper crankcase half checking that the front and rear balancer sprockets remain in alignment with the marked links on the chain. The remaining paint-marked link (9th) should align with the dot on the crankshaft sprocket. The balancer timing is correct when, at TDC, the paint-marked links of the balancer chain correspond with their respective timing marks and the front and rear balance weights are arranged so that their flat faces align with the arrows cast into the adjacent crankcase areas. If all is correct in this respect, the timing is set accurately and the crankcase halves can be joined.

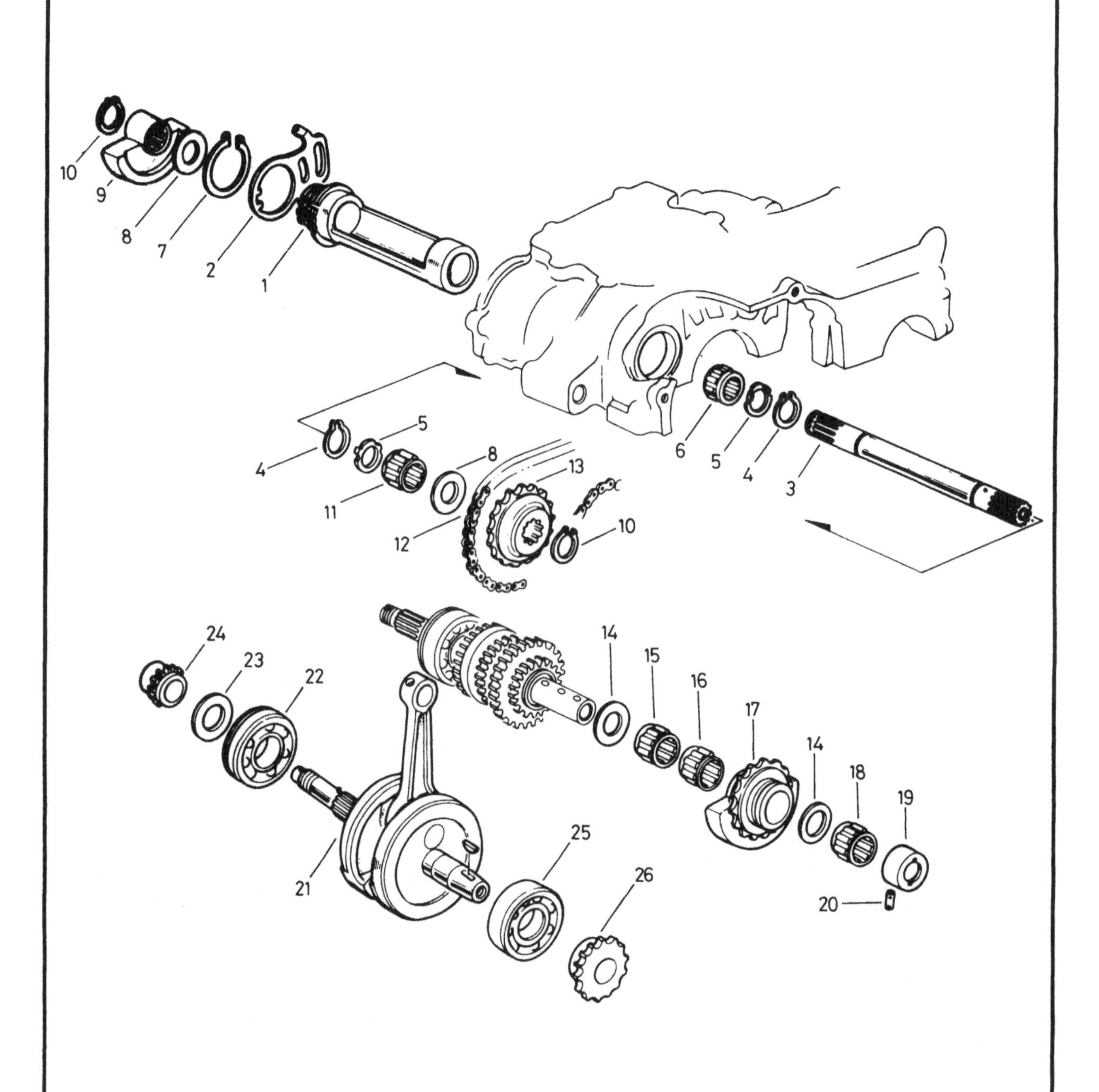

Fig. 1.13 Balancer mechanism and crankshaft

1 Balancer shaft holder
2 Adjuster flange
3 Front balancer shaft
4 Circlip
5 Special washer
6 Needle roller bearing
7 Circlip
8 Thrust washer
9 Front balance weight
10 Circlip
11 Needle roller bearing
12 Balancer chain
13 Front balancer sprocket
14 Thrust washer – 2 off
15 Needle roller bearing
16 Needle roller bearing
17 Rear balance weight
18 Needie roller bearing
19 Outer race
20 Locating dowel
21 Crankshaft
22 Main bearing
23 Washer
24 Cam chain sprocket
25 Main bearing
26 Balancer chain sprocket

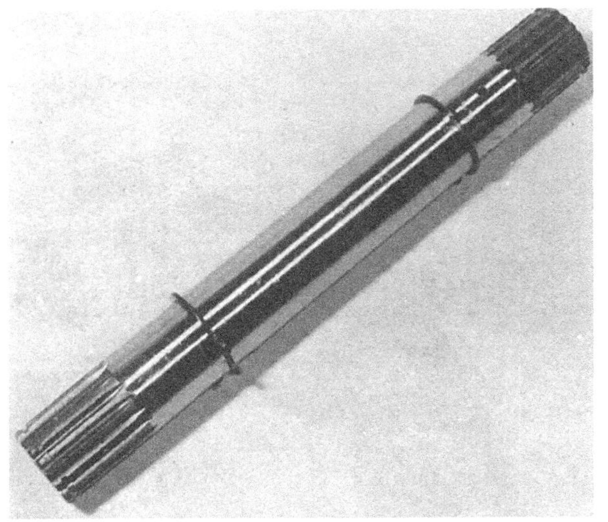

33.2a Fit circlips to grooves in front balancer shaft ...

33.2b ... and position the tanged thrust washers

33.3a Fit the needler roller bearing ...

33.3b ... followed by the large plain washer

33.3c Fit the balance weight to the splined shaft end ...

33.3d ... ensuring that timing marks align. Fit circlip

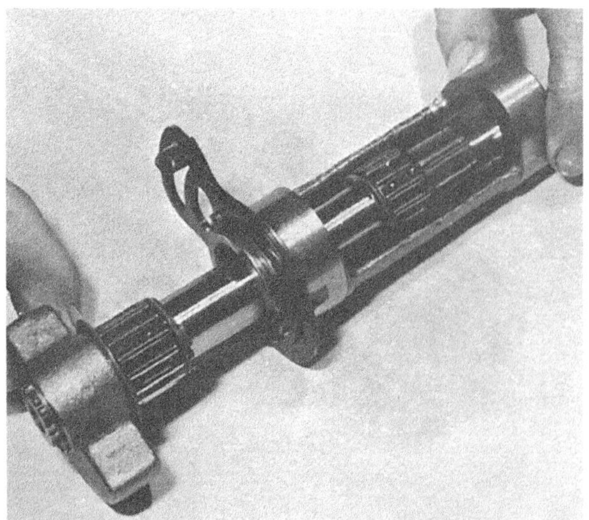

33.4 Slide the shaft and bearings into balancer holder

33.5 Balancer assembly is fitted in the direction shown

33.6a Rear balancer timing marks

33.6b Crankshaft sprocket timing marks

33.7a Front balancer timing marks

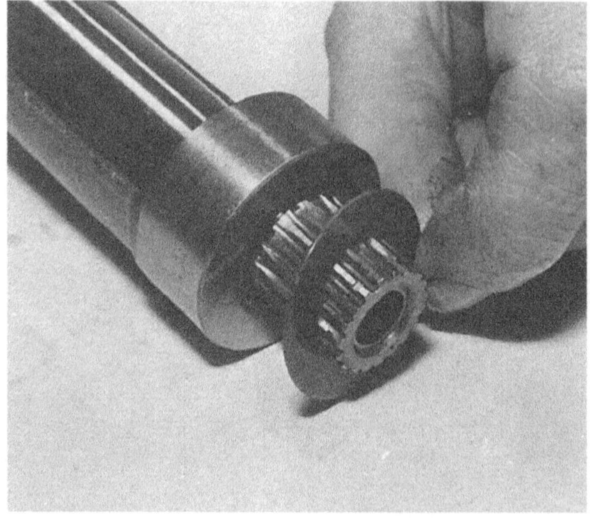

33.7b Note plain washer which precedes front sprocket

33.7c Note various timing marks on front assembly (arrowed)

33.7d Secure front sprocket with circlip

33.8a At TDC balancer should align with casting mark

33.8b ... as should the rear balancer

34 Engine reassembly: joining the crankcase halves

1 Having set up and checked the balancer timing as described in the preceding Section, the crankcase halves can be joined. The upper casing should be lifted clear and held or blocked in position while the mating surface of the lower crankcase is coated with a jointing compound. If essential, the balancer chain can be disengaged and the upper casing placed to one side, but do not forget that the timing will have to be re-checked if this course of action is chosen.

2 Apply a thin even film of jointing compound to the jointing face, using one of the silicone rubber based sealants. Take care not to get the compound in or near to any of the small oil passages. Leave the compound for a few minutes, then lower the upper casing half into position, ensuring that the balancer timing is preserved. Check that the two casing halves seat squarely, then fit the eight upper retaining bolts. These should be tightened in two stages to the torque setting specified below, following the tightening sequence shown in the accompanying diagram.

Upper crankcase bolt torque settings

6 mm bolts	*1.0 – 1.4 kgf m (7 – 10 lbf ft)*
8 mm bolts	*2.2 – 2.6 kgf m (16 – 19 lbf ft)*

3 Turn the unit over and fit the remaining crankcase bolts in a similar manner, tightening each one in two stages, following the accompanying sequence. The appropriate torque figures are as follows:

Lower crankcase bolt torque settings

6 mm bolts	*1.0 – 1.4 kgf m (7 – 10 lbf ft)*
8 mm bolts	*2.2 – 2.8 kgf m (16 – 20 lbf ft)*

It will be noted that the upper crankcase is secured by 6 mm bolts, with the exception of the rearmost left-hand bolt, which is 8 mm. On the underside of the unit, four 8 mm bolts are fitted in the centre, whilst five 6 mm bolts secure the front edge. Before moving on, refit the balancer chain guide block above the crankshaft on the left-hand side of the unit. The balancer chain tension can now be set as described in Section 35.

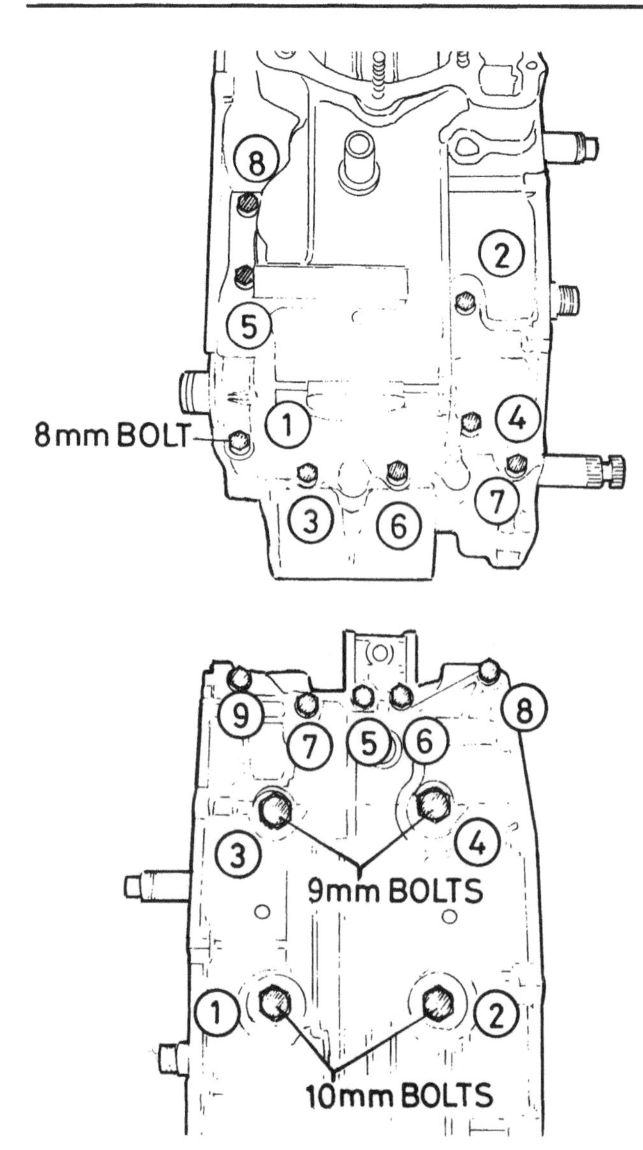

Fig. 1.14 Crankcase tightening sequence

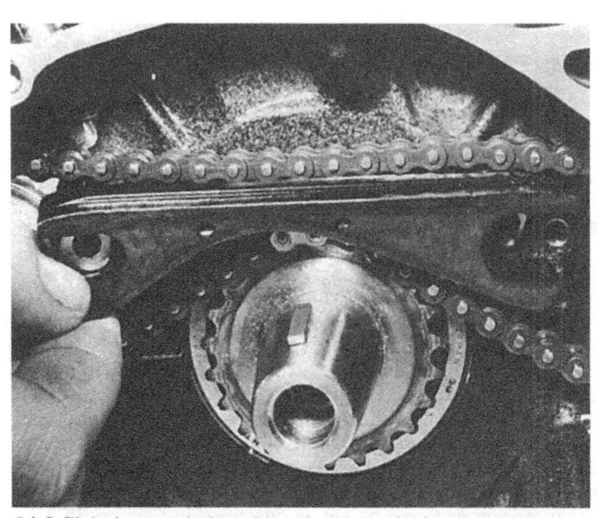

34.3 Fit balancer chain guide prior to tensioning

35 Engine reassembly: setting the balancer chain tension

1 Once the crankcase halves have been assembled, the balancer chain tension should be set to provide the correct operating conditions for the chain and sprockets. To achieve this, it is necessary to vary the distance between the front and rear balancers, and it is for this purpose that the front balancer holder is provided. The front balancer shaft runs in the eccentric bore of a tubular holder. The right-hand end of the holder terminates in an adjuster plate which is locked by a single retaining bolt. When adjustment is necessary the holder is rotated in the crankcase, effectively moving the axis of the front balancer shaft forwards or backwards.
2 To adjust the chain tension it will first be necessary to fit the adjuster spring to the adjuster plate or flange. Check that the holder moves freely in the casing and turns fully anti-clockwise under spring pressure. Check that the chain is held taut in this position.
3 The lower edge of the adjuster plate is marked by a series of lines. The balancer chain tension is set by moving the adjuster plate back (clockwise) by one graduation from the fully taut position. Holding this position, refit and tighten the locking bolt to 2.0 – 2.6 kgf m (14 – 19 lbf ft).
4 Occasionally it may prove impossible to obtain sufficient adjustment within the range of movement provided by the elongated slot. If this proves to be the case it will be necessary to move the adjuster plate in relation to the holder. Disconnect the adjuster spring. Release the circlip which retains the balance weight, then slide the latter off the end of the balancer shaft. Remove the plain washer which is fitted behind the balancer weight, then release the large circlip which secures the adjuster plate to the holder. It will be seen that the adjuster plate is located by tangs which engage in corresponding slots in the holder. Withdraw the adjuster plate and reposition it one slot further round (clockwise) to bring the range of adjustment within the scope of the plate.
5 Reassemble the balancer components by reversing the dismantling sequence, noting that the balancer timing mark must align with its counterpart on the balancer shaft. The tensioning operation can now be completed as described above.

36 Engine reassembly: refitting the oil pump

1 The oil pump may be fitted at any stage of reassembly after the selector forks have been fitted, but prior to the installation of the clutch. If the pump was dismantled for overhaul ensure that it has been assembled correctly and that the alignment dowel is in position. The pump should be primed by introducing oil into the inlet orifice whilst the pump pinion is rotated. It should be noted that the pump cover and body are located by a dowel, the single cross-head securing screw and the projecting end of the selector shaft over which the pump is positioned. It will be seen that there is a possibility of mis-alignment if all three location points are not present when the screw is tightened. For this reason it is suggested that the pump is temporarily placed in position in the casing and the retaining screw tightened. Remove the pump and check that it turns freely and evenly before final installation.
2 Check that the casing is perfectly clean, then fit new O-rings to the pump inlet and outlet orifices. Lower the pump into position over the projecting end of the selector fork shaft making sure that it seats squarely on the O-ring in the casing recess. Place the oil pump idler pinion in position over the selector fork shaft end, then offer up the retainer plate with the two, retaining bolts. It is essential that the plate engages in the recess in the end of the selector fork shaft because it serves to locate the shaft in the correct position to ensure an adequate oil supply to the gearbox. When everything is in position the two retaining bolts can be tightened down evenly.

35.2 Fit tensioner spring and set chain tension

36.2 Use new O-rings at oil pump ports

37 Engine reassembly: rebuilding and installing the clutch

1 Clutch assembly is facilitated by building the plates around the clutch centre, then fitting the assembly into the clutch drum. Start by identifying the outer friction plate which abuts the clutch centre flange. This has a larger internal diameter than the other friction plates, and thus a narrower friction area. Fit the outer friction plate together with the thin anti-judder seat (flat section) followed by the anti-judder spring (dished section). The three components can be retained by placing one of the plain clutch plates against them.
2 Continue assembly, fitting friction and plain plates alternately. Note that where new friction plates are fitted, these should be coated with engine oil prior to assembly. Fit the clutch pressure plate, passing the four projecting pillars through the corresponding holes in the clutch centre. The projecting tangs of the friction plates should be moved as required to bring them into alignment. This is important to ensure that the assembly will slide easily into the clutch outer drum.
3 Place the large plain thrust washer into its recess in the centre of the clutch drum, followed by the clutch plates and centre assembly. It will be necessary to lock the clutch to allow the centre nut to be secured, using the same method as was employed during removal. Place two of the clutch springs in position, securing them with two bolts and plain washers. Fit the flanged thrust washer, with the flanged head inwards, over the end of the gearbox mainshaft, followed by the clutch centre bush. Slide the assembled clutch into position.
4 Place the dished locking washer over the mainshaft end, then fit and secure the clutch centre nut tightening it to 4.0 – 6.0 kgf m (33 – 43 lbf ft). It will be necessary to prevent the clutch from rotating by whatever means was used during the dismantling sequence. Remove the two bolts and plain washers and refit the clutch release plate and springs, tightening the four retaining bolts securely.
5 Make a final check of the kickstart assembly to ensure that the decompressor cam index marks are aligned, then place the thrust washer over the end of the kickstart shaft.

38 Engine reassembly: refitting the cam chain, tensioner, primary drive pinion and ignition rotor

1 Place the cam chain around the crankshaft sprocket, using the mark made during removal as reference to ensure that the chain faces in its original direction of rotation. As mentioned previously, a part-worn chain will cause rapid wear of the sprockets and excessive noise if refitted incorrectly. Needless to say, this does not apply if a new chain is used. The rest of the chain should be fed through the cam chain tunnel and rested against the crankcase. Fit the small oil feed pipe which runs across the centre of the tunnel.
2 Pass the cam chain tensioner down through the tunnel, and secure it with its pivot bolt at the lower end. Note the headed sleeve which passes through the tensioner eye and provides a bearing surface for the pivot bolt. The flanged end of the sleeve must face the crankcase.
3 The primary drive pinion should be fitted next. Note that the gear teeth are heavily chamfered on one side only, and these should face the crankcase when the pinion is positioned. Ensure that the cutaway section of the splines coincides with the small dowel pin in the crankshaft end.
4 Slide the ignition pickup rotor over the crankshaft end, turning it until the small notch in the rotor engages with the dowel pin in the crankshaft. If a new rotor or stator assembly has been fitted, check carefully to ensure that each carries a matching identification mark. If the two are mis-matched, ignition performance will suffer. Fit the oil feed pad (quill) and spring into the crankshaft end, retaining them with the small securing pin. Fit the washer and securing nut, tightening the latter to 4.5 – 6.0 kgf m (33 – 43 lbf ft). Check that the rotor moves freely against spring pressure, returning fully when released. If it is not already in position, fit the oil filter screen.

37.1 Fit anti-judder spring seat, spring and special friction plate

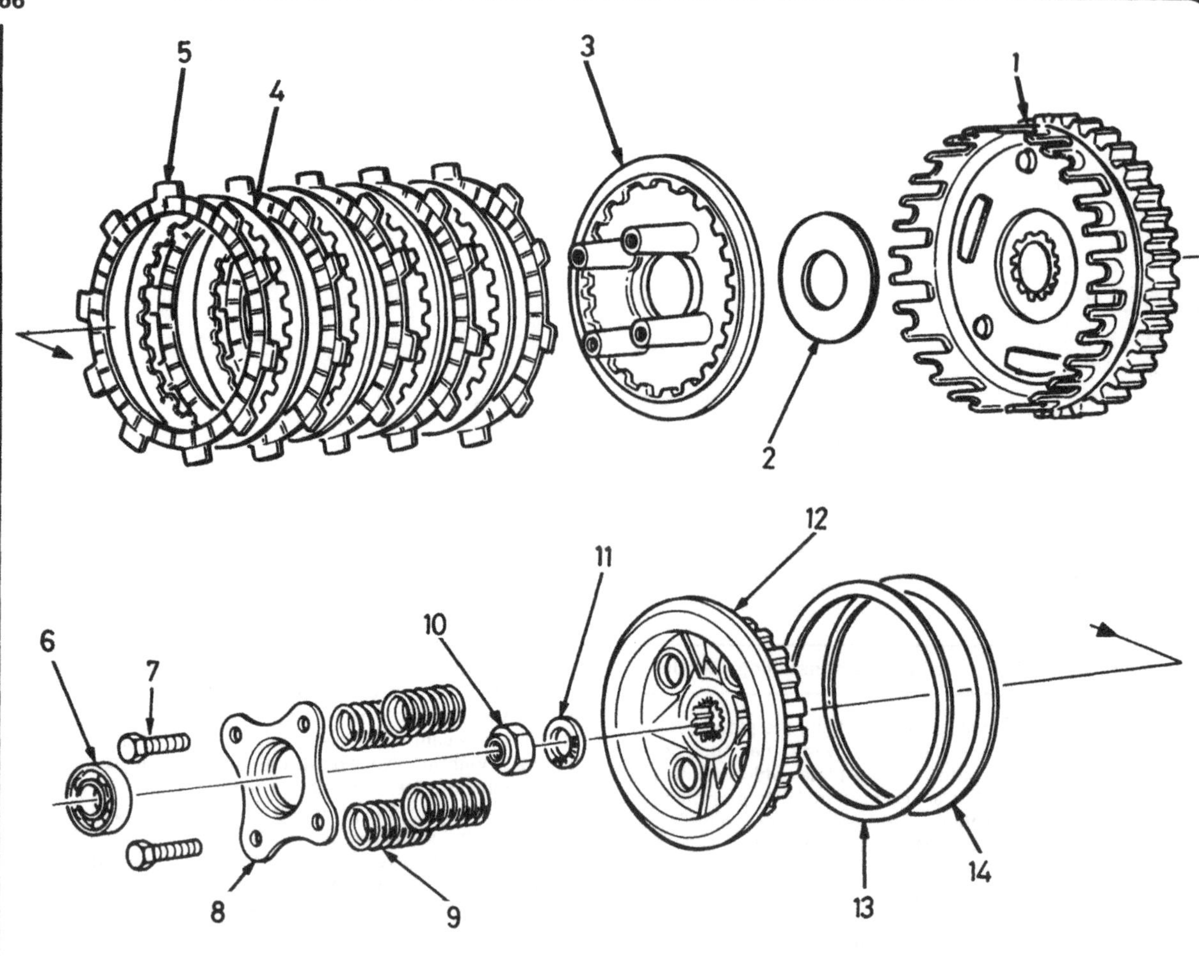

Fig. 1.15 Clutch

1 Outer drum
2 Thrust washer
3 Pressure plate
4 Plain plate – 4 off
5 Friction plate – 5 off
6 Thrust bearing
7 Bolt – 4 off
8 Clutch release plate
9 Spring – 4 off
10 Nut
11 Belville washer
12 Clutch centre
13 Anti-judder seat
14 Anti-judder spring

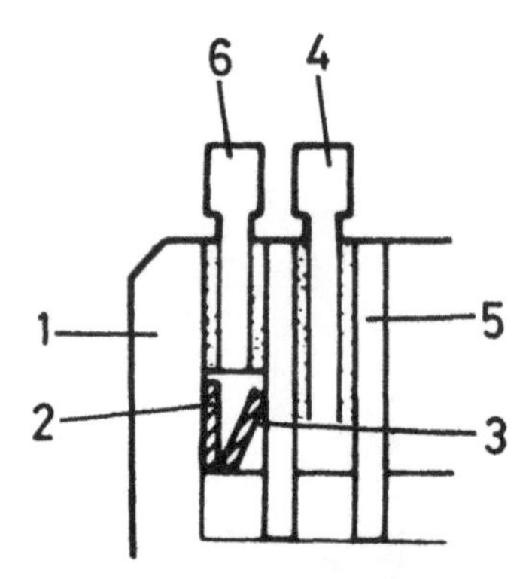

Fig. 1.16 Cross section of clutch judder spring and sea when correctly positioned

1 Clutch centre
2 Anti-judder seat
3 Anti-judder spring
4 Friction plate
5 Plain plate
6 Special inner friction plate

37.2a Assemble clutch plates over clutch centre ...

37.2b ... then fit the clutch pressure plate

37.3a Place thrust washer in clutch drum recess ...

37.3b ... and slide clutch centre assembly into place

37.3c Check that stepped spacer and bush are in place

37.4a Note direction marking on locking washer

37.4b Fit clutch assembly and lock plates as shown

37.4c Tighten centre nut to specified torque figure

37.4d The release plate can now be fitted

37.5 Check that cam is still in place and fit washer

38.1a Fit chain around crankshaft sprocket

38.1b Do not omit the oil feed pipe

38.2 Note position of shouldered spacer at tensioner mounting

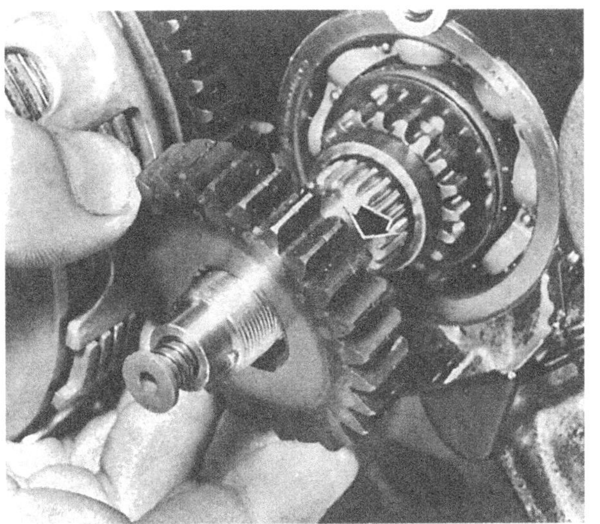

38.3 Primary drive pinion is fitted with chamfered teeth inwards (arrowed)

38.4a Fit automatic timing unit ...

38.4b ... and secure with nut

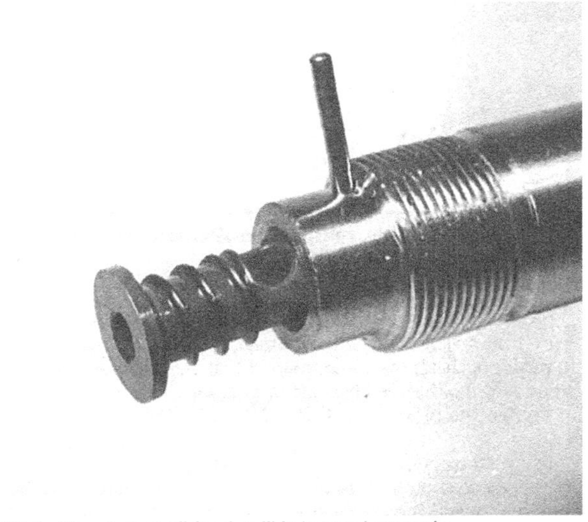

38.4c Check that oil feed quill is located properly

38.4d Tighten nut to specified torque setting

38.4e Check that filter screen is clean and refit

38.4f Outer cover can now be refitted

39.1a Fit Woodruff key and slide rotor into position

39.1b Lock crankshaft and tighten rotor bolt to specified torque

39 Engine reassembly: refitting the alternator and left-hand outer cover

1 Fit the Woodruff key into its slot in the crankshaft end and ensure that it seats fully and squarely. Slide the alternator rotor into position. Using the same method that was employed during dismantling, hold the crankshaft to prevent its rotation, then fit and tighten the retaining bolt to 10.0 - 12.0 kgf m (72 - 87 lbf ft).
2 If the stator assembly was removed from the inside of the outer cover it should be refitted in its original position. A line drawing showing the disposition of the stator, the spacer block and the locating dowels can be found in Chapter 6. Check that the gasket face of the cover and crankcase are clean and dry, then fit a new gasket. Note that a locating dowel is fitted near the front of the casing and a second dowel is fitted adjacent to the gearchange shaft. Offer up the cover, fitting and tightening the hexagon-headed screws to 0.8 - 1.2 kgf m (6 - 9 lbf ft). The two inspection caps in the cover should be left off at this stage to facilitate crankshaft rotation and timing checks.

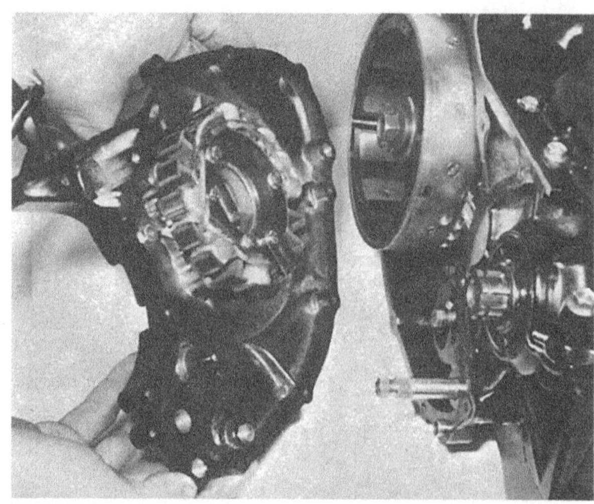
39.2 Fit dowels and gasket, then refit outer cover

40 Engine reassembly: refitting the piston, cylinder barrel and cylinder head

1 The above components can be fitted with the engine in or out of the frame. Commence reassembly by cleaning the mating faces of the cylinder head, barrel and crankcase to remove any residual dirt, oil or pieces of gasket. Check that the valves are installed correctly and that the necessary gaskets and O-rings are to hand before proceeding further.
2 The piston rings should be fitted to the piston, and it is important that the correct approach is adopted to avoid breakage of the very brittle rings. Starting with the lower (oil) ring install the convoluted expander section, ensuring that the ends butt together and do not overlap or become tangled. The two steel rails, which fit either side of the expander, are fairly flexible, and do not snap easily. They should be worked into position, taking care to position the end of each rail about 30° from the ends of the expander.

3 The two compession rings differ in section, and each must be fitted in its appropriate groove. The 2nd or middle ring has a tapered working face. This must face upwards, as should the letter 'N' which is etched on the top surface of each ring, adjacent to the ring end gap. The top or 1st ring can be identified by its plain section.
4 With a certain amount of practice, the ring gap can be spread sufficiently for the ring to be slipped into position. A safer method is to use three thin tin strips, spaced around the piston. These will allow the ring to be slid in position without fear of breakage. Arrange the ring gaps about 180° apart, and oil the rings and piston liberally.
5 Offer up the piston to the connecting rod ensuring that the IN mark on the piston crown faces rearwards. The gudgeon pin should push through fairly easily, but if it should prove stubborn, a rag soaked in hot water can be wrapped round the piston. The resulting expansion will allow the pin to be pushed through with ease. Retain the gudgeon pin with two new circlips. Resist the temptation to re-use old circlips, as they are invariably weakened during removal, and can become displaced during use, causing extensive (and expensive) engine damage. Note that the crankcase mouth should be packed with rag whilst the piston is installed, to prevent the ingress of debris or a displaced circlip.
6 Check that the cylinder barrel is clean, and that the oilways are clear. Use compressed air to blow the oilways out, or failing this, pipe cleaners or a strip of lint-free rag. Fit the two dowel pins which locate the cylinder barrel, noting that they fit around the front right-hand and rear left-hand studs. Fit a new cylinder base gasket to the crankcase, and a new O-ring around the projecting cylinder barrel liner.
7 Turn the crankshaft until the piston is at TDC, supporting the loop of camshaft chain to prevent it bunching against the casing. Coat the piston and cylinder bore with oil to aid assembly. Fitting the barrel unaided requires a certain amount of skill, and it may be helpful to have an assistant who can lower the barrel whilst the piston and rings are guided into place. The lead in or chamfer around the bottom of the cylinder liner makes assembly much easier. Lower the barrel slowly, compressing each ring in turn as it enters the bore. It should be noted that a piston ring compressor can be utilised, and this can often prove invaluable if the job is being done unaided. When the cylinder barrel spigot has engaged all the rings remove the rag padding from the crankcase mouth.
8 As the cylinder barrel is lowered into position, pull the camshaft chain up through the tunnel. Check that the cylinder barrel seats squarely, then fit the two flanged bolts which retain the barrel on the cam chain tunnel (right-hand) side. Do not tighten the bolts fully at this stage.
9 Pass a length of wire through the cam chain and secure it by passing it over the projecting tensioner. Lower the cam chain guide into the tunnel, ensuring that it locates in the recess in the cylinder barrel. Secure the cam chain tensioner assembly by fitting the sealing washer and lock bolt. If it has not already been locked in place, depress the tensioner to relieve tension and retain it by passing a pin through the 2 mm holding hole.
10 Fit the three cylinder head location dowels in their appropriately sized holes, noting the oil seal around the rear right-hand dowel, then fit a new cylinder head gasket. Lower the cylinder head into position, guiding the chain and tensioner through the aperture on the right-hand side. Fit the cylinder head bolts, tightening them evenly and progressively in a diagonal pattern to the following torque setting.

Cylinder head torque settings

Cylinder head bolts	*3.0 – 3.6 kgf m (22 – 26 lbf ft)*
Cylinder head nuts	*2.2 – 2.8 kgf m (16 – 20 lbf ft)*

Tighten the cylinder barrel bolts to 0.8 – 1.2 kgf m (6 – 9 lbf ft) then recheck the cylinder head nut torque settings. Fit the remaining tensioner lock bolt, tightening it to 0.8 – 12 kgf m (6 – 9 lbf ft).

40.5a Note "IN" mark on piston should face inlet port

40.5b Pack crankcase mouth with rag to catch errant circlips

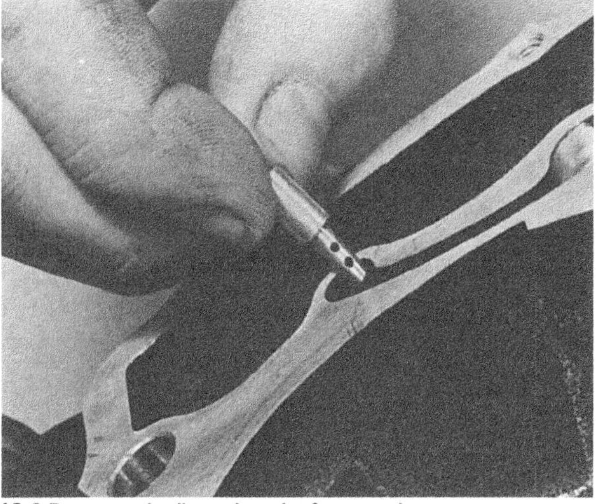
40.6 Do not omit oil restrictor jet from crankcase

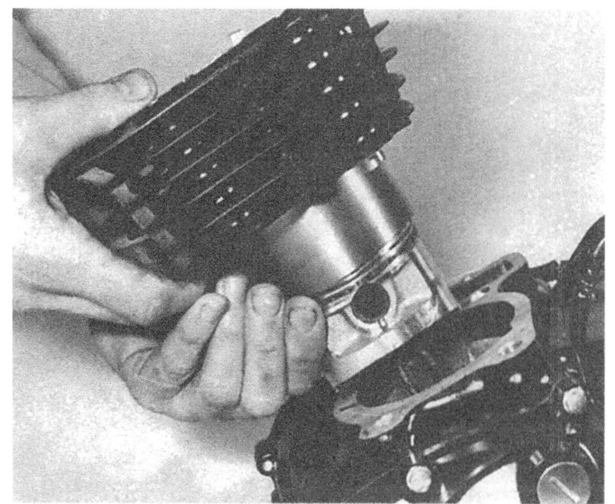

40.7 Taper in bore aids fitting of piston rings

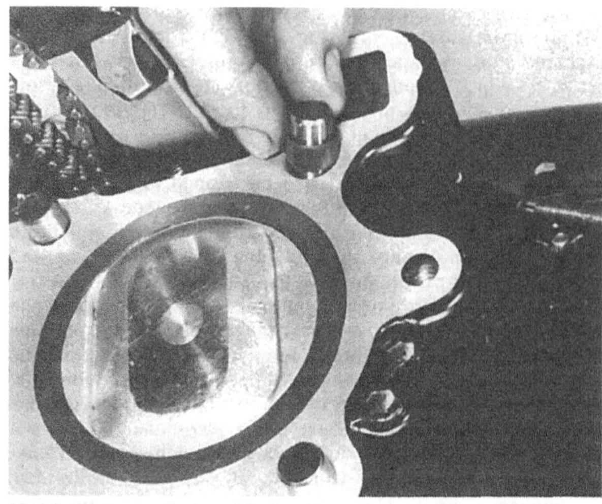

40.10a Note seal around hollow dowel pin

40.10b Fit new cylinder head gasket ...

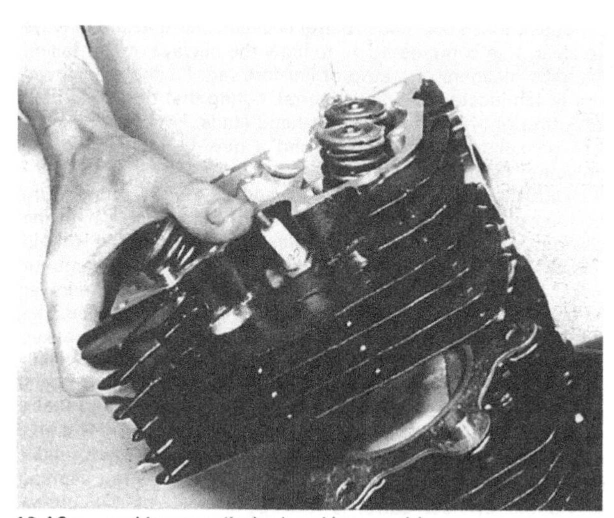

40.10c ... and lower cylinder head into position

40.10d Tighten head bolts in a diagonal sequence

41 Engine reassembly: refitting the camshaft and cylinder head cover – setting the valve timing

1 Check that the cylinder head area is clean and dry, then lubricate the cam bearing surface with molybdenum disulphide grease. Slide the camshaft into position from the right-hand side of the engine unit, looping the cam chain inside the sprocket flange. Offer up the camshaft sprocket with the two timing marks facing inwards, or towards the sparking plug.

2 Using a socket or box spanner passed through the access hole in the left-hand outer casing, turn the crankshaft anti-clockwise until the 'T' mark aligns with the index mark in the upper inspection hole. Arrange the camshaft sprocket so that the timing marks run parallel to the cylinder head gasket face, then place the cam chain around the sprocket without rotating either the sprocket or the crankshaft.

3 Arrange the camshaft so that the lobes point in roughly the 4 o'clock and 8 o'clock positions. The sprocket mounting holes should now coincide and the first of the two retaining bolts can be fitted. Turn the crankshaft through one complete revolution until the second bolt can be fitted. Tighten each one to 1.7 – 2.3 kgf m (12 – 16 lbf ft). Check the timing accuracy by turning the crankshaft, again anti-clockwise, until the 'T' mark aligns. Check that the cam sprocket timing marks align. Remove the locking

pin to release the cam chain tensioner. The tensioner should now automatically move to take up chain tension. If it appears to be stuck, push down on wedge B to allow wedge A to assume normal operation.

4 If the rocker arms and spindles were removed from the cylinder head cover, they should be refitted, lubricating the spindles and rocker arm bores with engine oil during assembly. Ensure that each rocker shaft is fitted in its original position and that the wave washer is positioned at the sparking plug end of each rocker. When fitting the retaining pins, position each shaft so that the cutout coincides with the stud hole through the cylinder head cover. The shaft can be turned as required using the slot provided in the shaft end. New O-rings should be fitted to obviate any risk of oil leakage.

5 The jointing faces of the cylinder head cover and the cylinder head should be clean and dry, and the cylinder head face coated with a thin, even film of silicone rubber jointing compound. Care must be exercised when applying the compound to avoid any excess reaching the camshaft where it might obstruct oilways. To this end, the compound should not be applied to the jointing face immediately around the two bearing journals particularly on the two pillars of the centre bearing.

6 Fit the two locating dowels to the cylinder head, then lubricate the valves and camshaft with engine oil, filling the pocket in which the cam lobes run. Slacken the valve adjuster locknuts and screws, then fit the cylinder head cover. Fit the retaining bolts, noting that three of these must be fitted with copper sealing washers (see accompanying photograph). The original washers may be re-used if they are undamaged. Run the bolts up finger tight, then commence tightening in a criss-cross pattern to pull the cover down evenly, thus avoiding any risk of warpage. Tighten to the appropriate torque setting.

Cylinder head cover bolt torque settings
1.0 – 1.4 kgf m (7 – 10 lbf ft)
Note: bolts with copper washers should be tightened to: 1.0 – 1.2 kgf m (7 – 9 lbf ft)

7 Before refitting the inspection covers, adjust the valve clearances as follows. Turn the crankshaft anti-clockwise until the T mark on the alternator rotor registers with the fixed index mark in the inspection hole. Using feeler gauges of the appropriate size, set each adjuster to give the required clearance between it and the valve stem. When set correctly the feeler gauge should be a light sliding fit between the two components. The recommended valve clearances are as follows:

Valve clearances (cold engine)

Inlet	*0.05 mm (0.002 in)*
Exhaust	*0.10 mm (0.004 in)*

When the valve clearances have been set, the inspection cover can be refitted, using new O-rings where necessary.

41.1 Note alignment marks on inner face of sprocket

41.3a Check timing and then secure sprocket to camshaft

41.3b Lubricate camshaft and valves before fitting cover

41.6 Arrowed bolts must have copper sealing washers fitted

41.7 Check valve clearances, then fit inspection covers

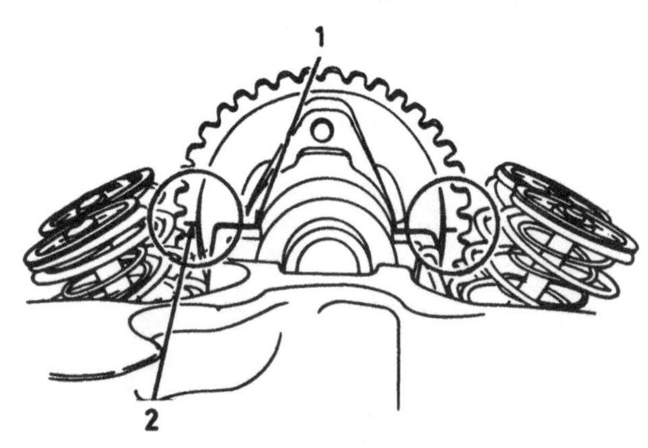

Fig. 1.17 Valve timing alignment marks

1 Cylinder head face
2 Sprocket alignment mark

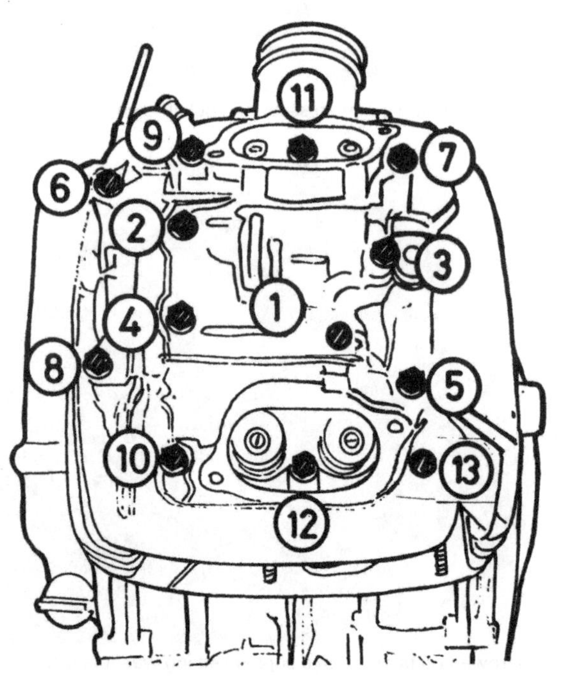

Fig. 1.18 Cylinder head tightening sequence

42 Engine reassembly: installing the rebuilt unit in the frame – final adjustments

1 As was the case during engine removal, installation can be completed by one person without insurmountable difficulties being encountered. It is helpful, however, to have some assistance whilst the unit is being lifted into position, as there will be less likelihood of the frame or paintwork sustaining damage. Lift the unit into the frame and assemble the front engine plates, the various mounting bolts, noting their varying lengths, and the cylinder head steady. Do not tighten any of the bolts until all are positioned loosely. The position of each bolt will be self evident and is indicated by its length. A mixture of 8 mm and 10 mm bolts are used, the frame to engine plate connections being made by the smaller 8 mm bolts whilst 10 mm bolts are employed in all other instances. When all of the bolts are in position they should be tightened to the appropriate torque settings shown below.

Engine mounting bolt torque settings
8 mm bolts 2.0 – 3.5 kgf m (14 – 25 lbf ft)
10 mm bolts 4.5 – 7.0 kgf m (51 – 72 lbf ft)

2 Refit the left-hand footrest assembly ensuring that the locating peg engages in its slot in the frame, and secure it with its single retaining bolt. Assemble the right-hand footrest and brake pedal unit in a similar fashion. Note the position of the combined brake pedal and kickstart stop plate which should be fitted as shown in the accompanying photograph. Reconnect the rear brake cable and brake switch operating spring.

3 Manoeuvre the carburettor into position, fitting the throttle and choke cables if these were released during removal. Set the lower cable adjusters to give approximately 2 – 6 mm of free play measured at the edge of the throttle twistgrip. Reconnect the choke cable, ensuring that it is set up so that the choke is fully off when the knob is pushed home. Set the carburettor vertically between the intake and air cleaner hoses, then tighten the retaining clips.

4 Place the gearbox sprocket in the loop of the final drive chain and slide the sprocket over the projecting splined end of the gearbox layshaft. Slide the retainer plate into position and turn it until the holes align with those in the sprocket. Coat the threads of the two securing bolts with a thread locking compound, then fit and tighten them. Check the tension of the final drive chain. This should be set to give 15 – 25 mm ($\frac{5}{8}$– 1 in) of vertical movement measured at the centre of the lower run. Check the brake pedal height, making any necessary adjustment using the stop bolt and locknut, then check the rear brake adjustment. This should be set to give 20 – 30 mm (0.8 – 1.2 in) free movement at the pedal end. Fit the pressed steel guard to the underside of the left-hand cover immediately below the gearbox sprocket, then refit the moulded plastic sprocket cover.

5 Connect the clutch cable to its operating arm, adjusting the cable to give 15 – 25 mm ($\frac{5}{8}$ – 1 in) free play measured at the lever end. Refit the decompressor cable between the cylinder head cover and the right-hand outer cover. The cable should be adjusted as described below noting that this operation must be carried out **after** the valve clearances have been set.

6 Using the inspection hole provided in the centre of the left-hand outer cover, turn the crankshaft anti-clockwise until the engine is on the compression stroke. This will be indicated by a high resistance to movement being felt. Carry on turning the crankshaft until TDC (top dead centre) is reached, this being indicated in the upper inspection hole by the T mark coinciding with the index mark. Check the amount of free play at the tip of the decompressor lever on the cylinder head cover. This should be set using the cable adjuster to give 1 – 3 mm (0.04 – 0.12 in) free play.

7 Trace the ignition and alternator leads, routing them along the frame tubes to their corresponding connector halves. Once connected, the cables should be secured to the frame tubes

using new cable ties, or in an emergency, pvc tape. Assemble the exhaust pipes using new exhaust port sealing rings to prevent leakage. Fit the split collets and retainers at the exhaust ports. The integral silencer section of each exhaust system is secured by the pillion footrest assembly. Note that the silencer bracket has a slot that locates a tang on the footrest bracket.

8 Assemble the gearchange linkage as shown in the accompanying photograph. It is important that the assembly is positioned in approximately the correct position. Check that the pedal end is conveniently positioned in relation to the footrest and that gearchanges are possible without the linkage becoming fouled.

9 Refit the battery ensuring that the battery leads are not inadvertently reversed. Turn on the ignition and check that the electrical system functions normally. Refit the fuel tank, seat and side panels, noting that the tank mounting rubbers may be lubricated with aerosol maintenance spray or petrol to aid installation. Reconnect the feed pipe between the carburettor and the fuel tank.

10 Remove the combined crankcase filler/dipstick and add 1.7 litres (3.0 Imp pints) of SAE 10W-40 motor oil. Hold the machine upright and rest the combined filler plug and dipstick on the top of the filler hole. Withdraw the dipstick and check that the oil level is within the upper and lower limits. Note that the engine oil level must be rechecked soon after the engine has been run and then left to settle for a while because a small amount of topping up may be necessary after the oil has been distributed around the engine and transmission components.

43 Starting and running the rebuilt engine

1 Turn on the fuel tap, close the choke, and attempt to start the engine by means of the kickstart pedal. Do not be disillusioned if there is no sign of life initially. A certain amount of perseverance may prove necessary to coax the engine into activity even if new parts have not been fitted. Should the engine persist in not starting, check that the sparking plug has not become fouled by the oil used during re-assembly. Failing this go through the fault finding charts and work out what the problem is methodically.

2 When the engine does start, keep it running as slowly as possible to allow the oil to circulate. Open the choke as soon as the engine will run without it. During the initial running, a certain amount of smoke may be in evidence due to the oil used in the reassembly sequence being burnt away. The resulting smoke should gradually subside.

3 Check the engine for blowing gaskets and oil leaks. Before using the machine on the road, check that all the gears select properly, and that the controls function correctly.

42.2 Note position of brake pedal stopper plate

42.3a Refit carburettor to mounting stubs

42.3b Refit adaptor mounting bolts, where necessary

42.3c Connect the throttle and choke cables

42.4 Refit sprocket and retainer and tighten bolts

42.5 Fit clutch cable anchor plate and connect cable

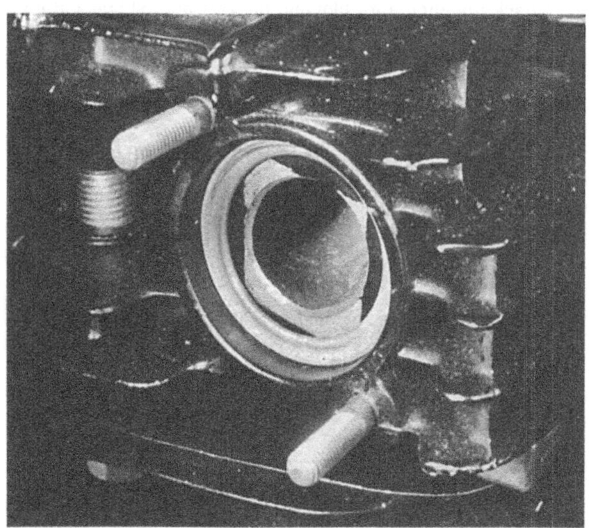

42.7a Use new sealing rings in exhaust ports ...

42.7b ... and assemble retainers as shown

42.8 Note position of gearchange linkage

42.9a Tank rubbers retain cable guides

42.9b Reconnect carburettor feed and drain hoses

42.10 Top up crankcase with engine oil

44 Fault diagnosis: engine

Symptom	Cause	Remedy
Engine will not start	Defective sparking plug	Remove the plug and lay it on cylinder head. Check whether spark occurs when ignition is switched on and engine rotated. If no spark occurs fit replacement plug.
	Faulty ignition pulser of CDI unit	See Chapter 3.
	Fuel system fault	Slacken carburettor drain screw to check whether fuel is reaching the float bowl.
Engine runs unevenly	Ignition and/or fuel system fault	Check each system independently, as though engine will not start.
	Blowing cylinder head gasket	Leak should be evident from oil leakage where gas escapes.
	Incorrect ignition timing	See Chapter 3.
Lack of power	Fault in fuel system or incorrect ignition timing	See above.
Heavy oil consumption	Cylinder barrel in need of rebore	Check for bore wear, rebore and fit oversize piston if required.
	Damaged oil seals	Check engine for oil leaks.
Excessive mechanical noise	Worn cylinder barrel (piston slap)	Rebore and fit oversize piston.
	Worn big end bearing (knock)	Fit replacement crankshaft assembly.
	Worn main bearings (rumble)	Fit new journal bearings and seals.
Engine overheats and fades	Lubrication failure	Stop engine and check whether internal parts are receiving oil. Check oil level in crankcase.

45 Fault diagnosis: clutch

Symptom	Cause	Remedy
Engine speed increases as shown by tachometer but machine does not respond	Clutch slip	Check clutch adjustment for free play at handlebar lever. Check thickness of inserted plates.
Difficulty in engaging gears. Gear changes jerky and machine creeps forward when clutch is withdrawn. Difficulty in selecting neutral	Clutch drag	Check clutch adjustment for too much free play. Check clutch drums for indentations in slots and clutch plates for burrs on tongues. Dress with file if damage not too great.
Clutch operation stiff	Damaged, trapped or frayed control cable	Check cable and replace if necessary. Make sure cable is lubricated and has no sharp bends.

46 Fault diagnosis: gearbox

Symptom	Cause	Remedy
Difficulty in engaging gears	Selector forks bent Gear clusters not assembled correctly	Replace. Check gear cluster arangement and position of thrust washers.
Machine jumps out of gear	Worn dogs on ends of gear pinions Stopper arms not seating correctly	Replace worn pinions. Remove right hand crankcase cover and check stopper arm action.
Gearchange lever does not return to original position	Broken return spring	Replace spring.
Kickstarter does not return when engine is turned over or started	Broken or poorly tensioned return spring	Replace spring or re-tension.
Kickstarter slips	Ratchet assembly worn	Part crankcase and replace all worn parts.
Engine hard to turn using kickstart	Decompressor inoperative or out of adjustment	Check and adjust.

Chapter 2 Fuel system and lubrication

For information relating to the CB250 RSD-C model, refer to Chapter 7

Contents

Specifications

Fuel capacity	
Overall	13 litre (2.9 Imp gal)
Reserve	3 litre (6.4 Imp pint)
Carburettor	
Make	Keihin
Type	PD 70A
Main jet	122
Slow jet	40
Jet needle clip position	4th groove
Pilot screw setting	$1\frac{3}{4}$ turns out
Accelerator pump delivery	0.15 cc per stroke
Idle speed	1200 ± 100 rpm
Fast idle speed	2000 – 2500 rpm
Venturi diameter	30 mm (1.18 in)
Throttle valve diameter	28 mm (1.10 in)
Float level	14.5 mm (0.57 in)
Lubrication system	
Type	Wet sump, pressure fed
Filter	Gauze strainer
Oil pump	
Type	Trochoidal
Inner/outer rotor clearance	0.15 mm (0.006 in)
Service limit	0.20 mm (0.008 in)
Outer rotor/body clearance	0.15 – 0.18 mm (0.006 – 0.007 in)
Service limit	0.25 mm (0.010 in)
Rotor end float	0.01 – 0.07 mm (0.0004 – 0.0028 in)
Service limit	0.12 mm (0.0047 in)

1 General description

The CB250 RS is equipped with a conventional lever-operated slide carburettor which incorporates a throttle-controlled accelerator pump. Fuel is fed by gravity from the fuel tank via a three position fuel tap. The tap positions are 'Off', 'On' and 'Reserve', the latter position providing an emergency supply of fuel and a warning that the fuel level has run low. Fuel flows from the tank to the carburettor float chamber where the level is kept constant by a float-operated valve.

The fuel/air mixture entering the engine is controlled by the throttle valve, the needle/needle jet arrangement and by the main and pilot jets. A rod-operated accelerator pump is brought into action every time the throttle is opened. This provides the rich mixture required for acceleration, but allows the overall jetting to be set weaker than normal thus improving overall fuel consumption.

A comprehensive force-fed lubrication system is employed. A trochoid oil pump draws oil from the sump via a small gauze strainer which traps any metal particles. At the pump outlet the oil splits into a number of circuits. One of these feeds oil through a spring-loaded quill into the right-hand end of the crankshaft. The oil is routed through the crankshaft and flywheel to the big-end bearing, the escaping oil providing splash lubrication for the small-end bearing and cylinder wall before returning to the sump.

A branch off the crankshaft feed supplies oil to the camshaft and rockers and also to the front balancer shaft. Oil is also fed under pressure to the gearbox layshaft and mainshaft assemblies.

2 Fuel tank: removal, examination and replacement

1 To gain access to the single tank-retaining bolt it is first necessary to release the seat. This is secured by a bolt on each side of the seat base.

2 The seat can now be lifted clear to reveal the single tank retaining bolt which passes through a lug into the frame. Slacken and remove the bolt, then lift the rear of the tank upwards. This will make access to the fuel pipe easier, allowing it to be prised off once the tap has been turned to the 'Off' position. Once removed, the tank can be pulled back to free the front mounting rubbers and lifted away.

3 If the tank is to be stored it should be placed in a safe place away from any area where fire is a hazard or where the paint finish may become damaged.

4 Any sign of fuel leakage should be dealt with promptly in view of the risk of fire or explosion should fuel drip onto the hot exhaust system. It is **not** recommended that the tank is repaired using welding or brazing techniques, because even a small amount of residual fuel vapour can result in a dangerous explosion. A more satisfactory alternative is to use one of the resin-based tank sealing compounds. These are designed to line the tank with a tough fuel-proof skin, sealing small holes or splits in the process. The suppliers of these products advertise regularly in the motorcycle press.

5 Tank fitting is a straightforward reversal of the removal sequence. If problems are encountered when attempting to engage the front mounting rubbers, a trace of petrol will act as a lubricant, easing assembly appreciably. After the fuel pipe has been refitted, turn the fuel on and check for leaks.

3 Petrol feed pipe: examination

1 The petrol feed pipe is made from thin walled synthetic rubber and is of the push-on type. It is necessary to replace the pipe only if it becomes hard or splits. It is unlikely that the retaining clips will need replacing due to fatigue as the main seal between the pipe and union is effected by an interference fit.

2 If the petrol pipe has been replaced with the transparent plastic type for any reason, look for signs of yellowing which indicates that the pipe is becoming brittle due to the plasticiser being leached out by the petrol. It is a sound precaution to renew a pipe when this occurs, as any subsequent breakage whilst in use will be almost impossible to repair.

Note: On no account should natural rubber tubing be used to carry petrol, even as a temporary measure. The petrol will dissolve the inner wall, causing blockages in the carburettor jets which will prove very difficult to remove.

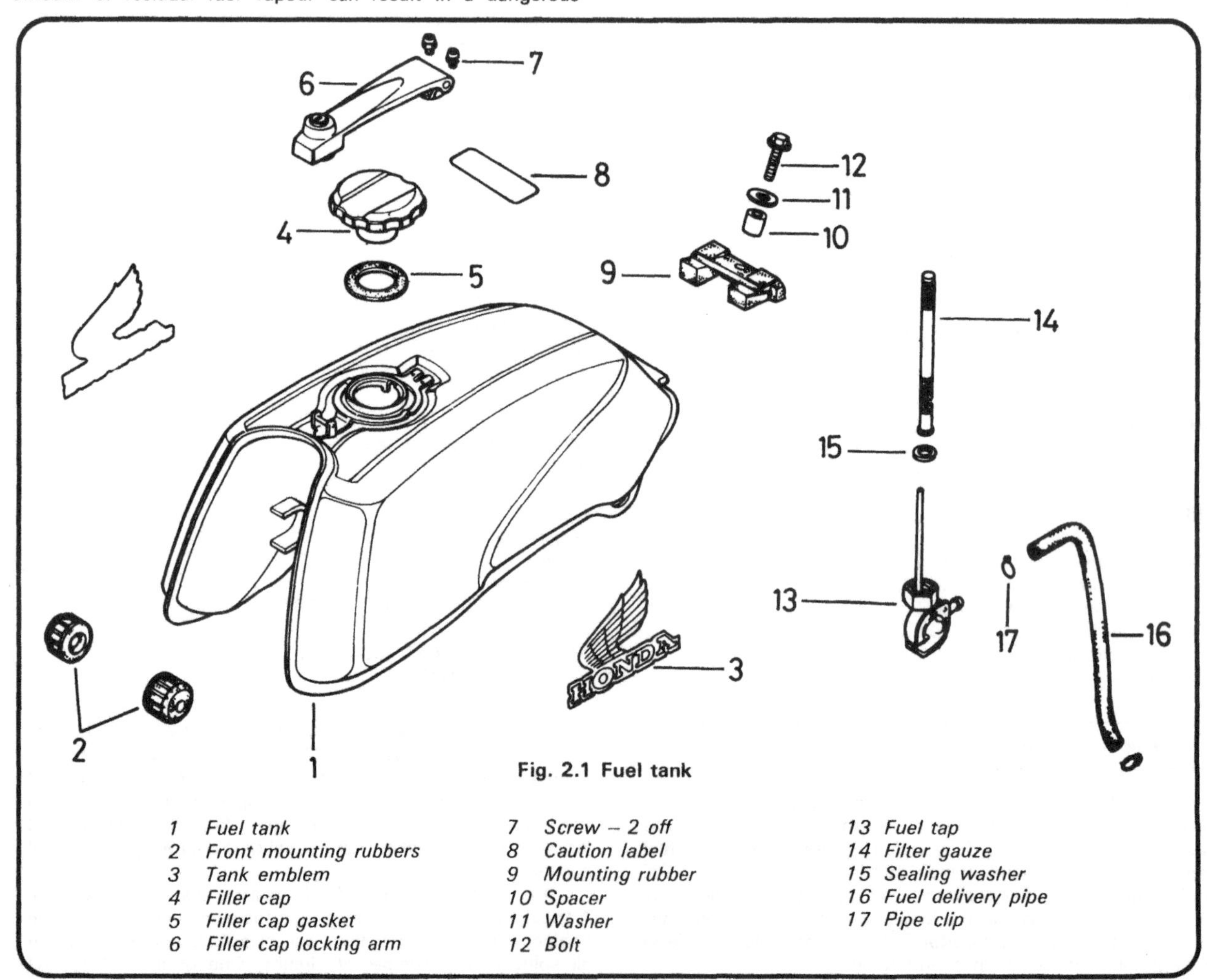

Fig. 2.1 Fuel tank

1 Fuel tank
2 Front mounting rubbers
3 Tank emblem
4 Filler cap
5 Filler cap gasket
6 Filler cap locking arm
7 Screw – 2 off
8 Caution label
9 Mounting rubber
10 Spacer
11 Washer
12 Bolt
13 Fuel tap
14 Filter gauze
15 Sealing washer
16 Fuel delivery pipe
17 Pipe clip

4 Petrol tap: removal, examination and replacement

1 Before the petrol tap can be removed, it is first necessary to drain the tank. This is easily accomplished by removing the feed pipe from the carburettor float chamber and allowing the contents of the tank to drain into a clean receptacle, with the tap turned to the 'Reserve' position. Alternatively, the tank can be removed and placed on one side, so that the fuel level is below the tap outlet. Take care not to damage the paintwork.
2 The tap unit is retained by a gland nut to the threaded stub on the underside of the tank. It can be removed after the fuel pipe has been pulled off the tap.
3 If the tap lever leaks, it will be necessary to renew the tap as a complete unit. It is not possible to dismantle the tap for repair.
4 When reassembling the tap, reverse the procedure for dismantling.
5 Check that the feed pipe from the tap to the carburettor is in good condition and that the push-on joints are a good fit, irrespective of the retaining wire clips. If particles of rubber are found in the filter, replace the pipe, since this is an indication that the internal bore is breaking up.
6 If there have been indications of water contamination in the fuel, the removal of the tap presents a good opportunity to drain and flush the tank completely. Many irritating fuel system faults can be traced to water in the petrol. This often appears as a result of condensation inside the petrol tank. The resulting blobs of water are easily drawn into the carburettor, where they can cause intermittent blockages in the jets and drillings. Any accumulations of water should therefore be flushed from the tank before the tap is refitted. The tubular filter gauze should be removed and cleaned carefully prior to reassembly.

5 Carburettor: removal

1 To gain access to the carburettor for removal purposes it is first necessary to remove the seat and fuel tank, as described in Section 2, and to pull off the side panels. Slacken the locknuts which retain the throttle cable adjusters to the anchor plate. The adjusters can now be displaced and the inner cables disengaged from the operating pulley. Slacken the choke cable clamping screw and release the choke cable from the carburettor.
2 Slacken the retaining clips which secure the carburettor to the inlet adaptor and the air cleaner hose. The carburettor can now be pulled back to free it from the adaptor and manoeuvre clear of the engine. If removal proves difficult it may be easier to release the two bolts which retain the intake adaptor to the cylinder head. The carburettor can be displaced with the adaptor attached.

6 Carburettor: dismantling and reassembly

1 Start by draining the residual fuel from the float bowl by means of the small drain screw which screws into the base of the float bowl. Slacken and remove the two screws which retain the carburettor top and lift the top away. Release the nut and washer on the end of the throttle spindle and remove the fast idle link, taking care to avoid straining the accelerator pump spring. Remove the small screw which retains the throttle link to the spindle.
2 Unhook the throttle return spring, and withdraw the quadrant and spindle. The throttle lever will now be freed and should be withdrawn together with the throttle valve assembly. The lever and throttle valve can be separated after releasing the connecting link which joins them. This is accomplished by removing the small tension spring which retains the pivot of the lever and the throttle valve bracket to the link. To complete the dismantling of the throttle valve assembly, remove the two small screws which secure the bracket to the throttle valve, then lift away the bracket, spring and jet needle.
3 The accelerator pump consists of a rod-operated diaphragm unit which is controlled by the throttle mechanism. The operating rod passes vertically upwards to a point just below the throttle lever pivot. A pump lever is connected by a spring to the fast idle link. The latter turns with the throttle pulley thus applying pressure to the pump lever. This in turn depresses the pump rod causing a metered quantity of fuel to be injected as the throttle is opened. To dismantle the pump components, release the three screws which retain the pump cover, lifting this and the return spring away. Carefully remove the pump diaphragm and withdraw it together with the pump rod. The pump lever pivots on a hexagon and cross-headed bolt and can be removed when this has been unscrewed.
4 Remove the three float bowl retaining screws, then lift the bowl away, taking care not to damage the sealing ring. The float can be removed after displacing and withdrawing its pivot pin. The float needle will come away with the float to which it is attached by a fine wire loop.
5 The main jet is screwed into the bottom of the needle jet holder which is screwed in turn into the centre of the carburettor body. The adjacent projection is the pilot (slow) jet which cannot be removed. The main jet, identified by its slotted cheese head, can be unscrewed separately from, or as a part of, the needle jet holder; the removal of the holder releases the needle jet which is pressed out from above by passing a wooden dowel down through the throttle valve bore. On removing the pilot mixture screw, screw it in until it seats lightly and record the exact number of turns required to do so before unscrewing it complete with its spring, washer and O-ring.
6 Reassembly is tackled in the reverse of the dismantling order, using new seals and O-rings as required. Each part must be scrupulously clean, and care must be exercised to avoid overtightening any of the carburettor components. All of these are rather delicate and can easily become damaged.
7 If disturbed during dismantling, the accelerator pump setting should be checked and if necessary adjusted to the specified clearance. With the throttle in the fully closed position measure the clearance between the top of the accelerator pump rod and the pump lever projection. This should be within the range of 0 – 0.04 mm (0 – 0.0016 in); if adjustment is required carefully bend the lever projection by a small amount to obtain the correct setting.
8 Refer to Routine maintenance for details of throttle and choke cable adjustment and the following sections for carburettor settings.

7 Carburettor: examination and renovation

1 Having dismantled the carburettor as described in Section 6, the various components should be laid out for examination. If symptoms of flooding have been in evidence, check that the float is not leaking, by shaking and listening for petrol inside. It is rare to find leaks in plastic floats, this problem being more common in the brass type.
2 A more likely cause of flooding is dirt on the float needle or its seat. Examine the faces of the needle and seat for foreign matter and also for scoring. If in bad condition, renew the needle and note whether any improvement is obtained. The valve seat cannot be removed from the body and if badly damaged the entire body must be renewed. The only alternative is to get the seat area refaced by a light engineering company, remembering that this will upset the float height setting.
3 The main jet screws into the needle jet holder, which is central in the carburettor body. It is not prone to any real degree of wear, but can become blocked by contaminants in the petrol. These can be cleared by an air jet, either from an air line or a foot pump. As a last resort, a fine bristle from a nailbrush or similar may be used, but on no account should wire be used as this may damage the precision drilling of the jet.
4 The needle jet may become worn after a considerable mileage has been covered and should be renewed along with the needle. Always fit replacement parts as a pair.
5 The pilot jet is located adjacent to the main jet and needle

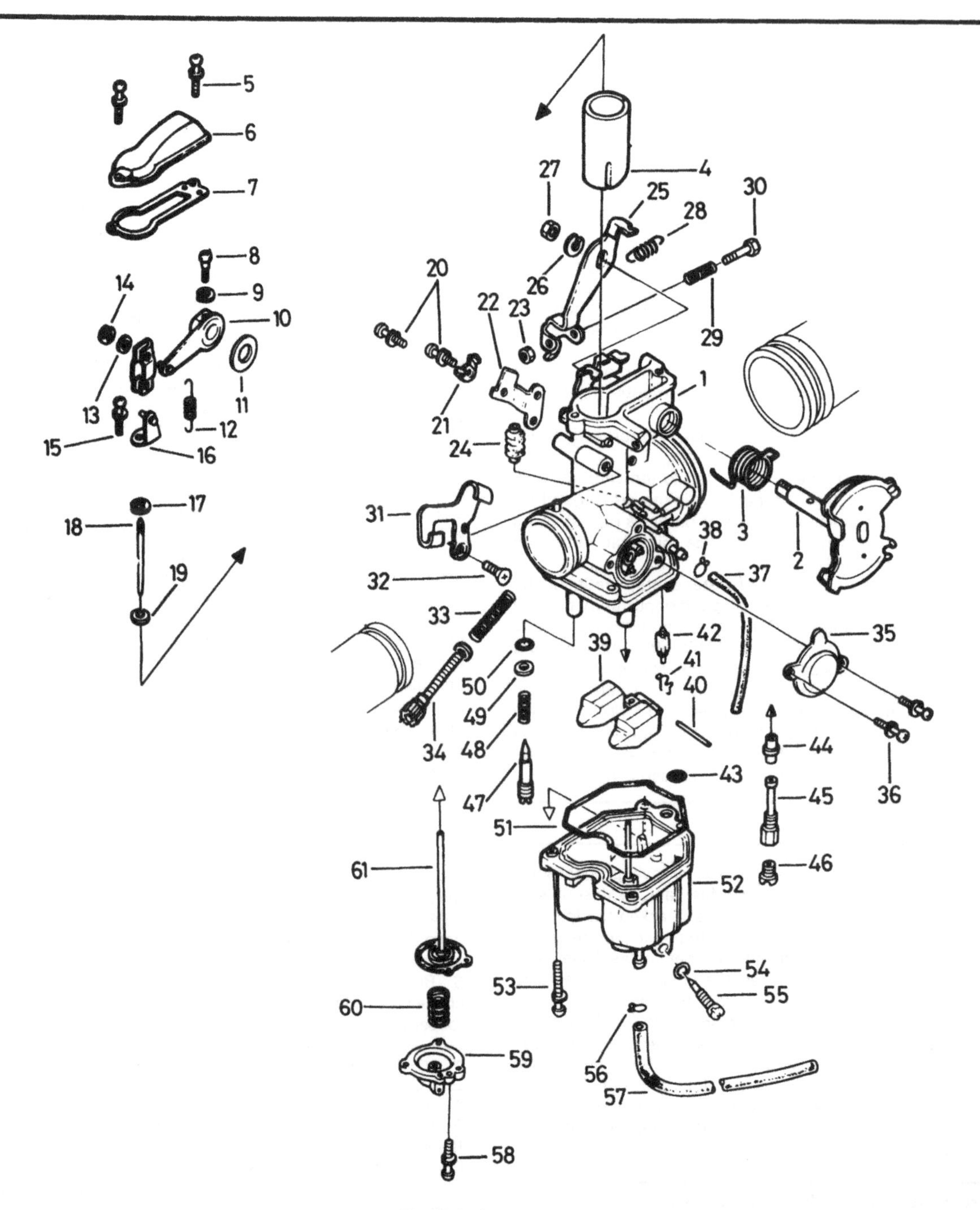

Fig. 2.2 Carburettor

1 Carburettor body
2 Throttle spindle/pulley
3 Return spring
4 Throttle valve
5 Screw – 2 off
6 Carburettor top
7 Gasket
8 Screw
9 Spring washer
10 Throttle lever
11 Washer
12 Tension spring
13 Washer
14 Circlip
15 Screw – 2 off
16 Throttle valve bracket
17 Needle clip
18 Jet needle
19 Washer
20 Screw
21 Cable clamp
22 Abutment bracket
23 Nut
24 Boot
25 Fast idle link
26 Spring washer
27 Nut
28 Spring
29 Spring
30 Bolt
31 Cable anchor bracket
32 Screw
33 Spring
34 Throttle stop screw
35 Blanking plate
36 Screw – 2 off
37 Vent hose
38 Pipe clip
39 Float
40 Float pivot pin
41 Fine wire loop
42 Float needle
43 O-ring
44 Needle jet
45 Needle jet holder
46 Main jet
47 Pilot mixture screw
48 Spring
49 Washer
50 O-ring
51 Float bowl sealing ring
52 Float bowl
53 Screw
54 Washer
55 Drain screw
56 Pipe clip
57 Overflow pipe
58 Screw – 3 off
59 Accelerator pump cover
60 Spring
61 Accelerator pump diaphragm

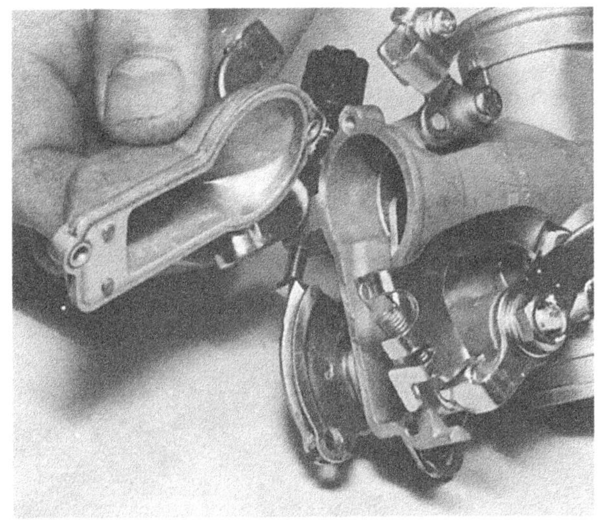
6.1a Carburettor top is retained by two screws

6.1b Disengage and remove the return spring

6.1c Slacken retaining nut on end of throttle spindle ...

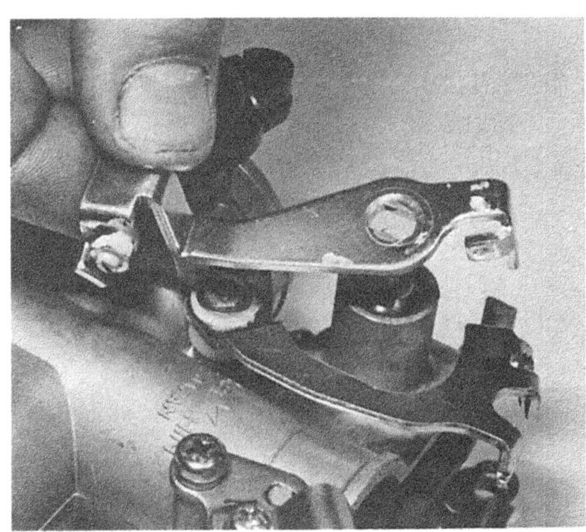
6.1d ... and lift the fast idle link away

6.1e Slacken the throttle link retaining screw

6.2a Disengage the throttle return spring and withdraw spindle

6.2b Link and throttle valve can now be withdrawn – note washer

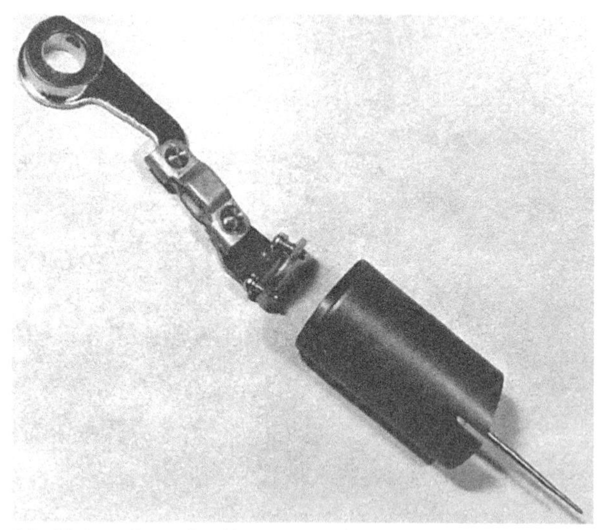

6.2c Linkage is retained by two screws

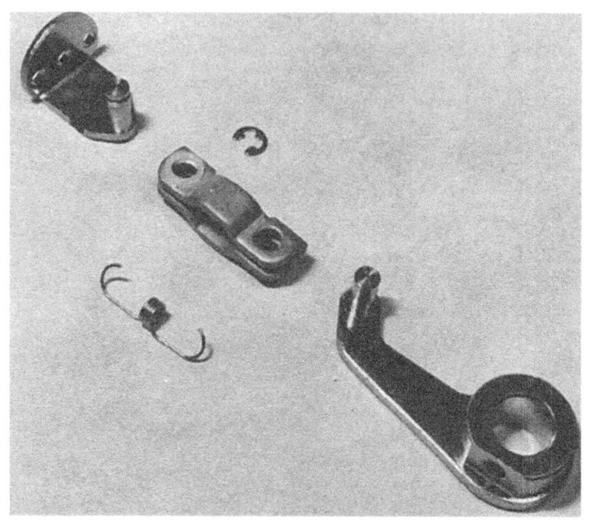

6.2d Operating linkage can be dismantled as shown

6.3a Slacken the pump cover securing screws ...

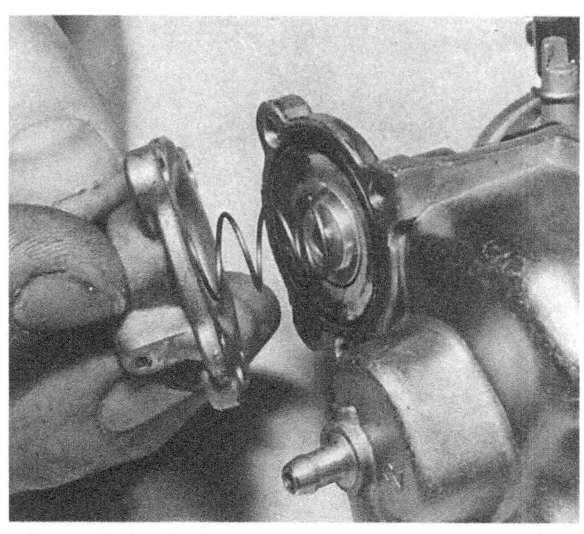

6.3b ... and remove the cover and return spring

6.3c Carefully peel away diaphragm edge and remove

6.3d Remaining cover serves as blanking plate only

6.4a Release screws and lift float bowl away

6.4b Use a piece of wire to displace float pivot pin

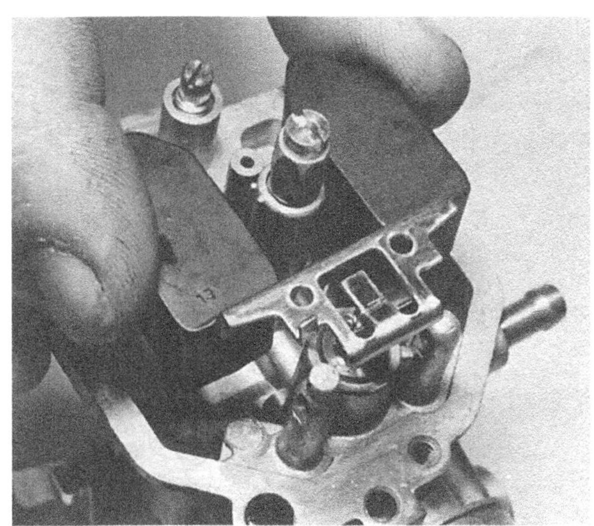
6.4c Float assembly can now be lifted away

6.4d Float needle is retained by wire clip

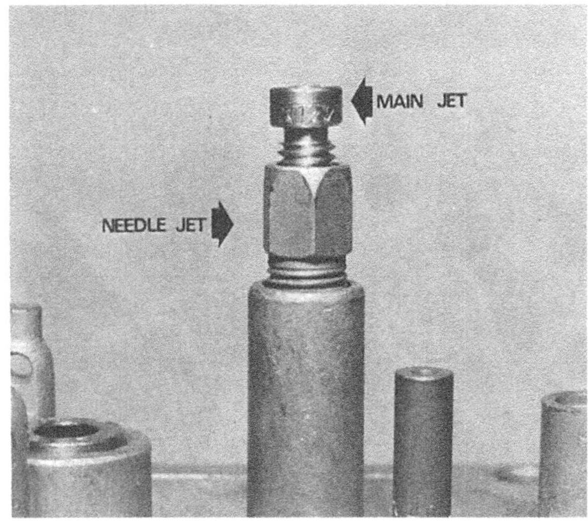

6.5a Main and needle jet location

6.5b Needle jet incorporates bleed holes

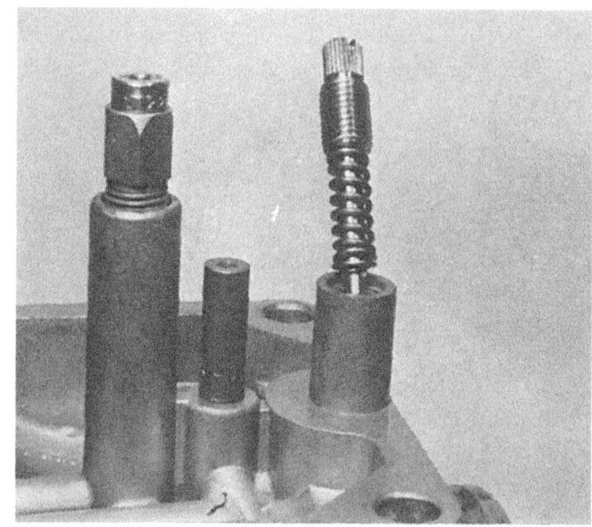
6.5c Note pilot screw setting before removal

6.6 Check gasket and O-ring and renew as required

7.1 Check float for leakage and valve needle for wear

7.2 Valve seat (arrowed) is pressed into carburettor body

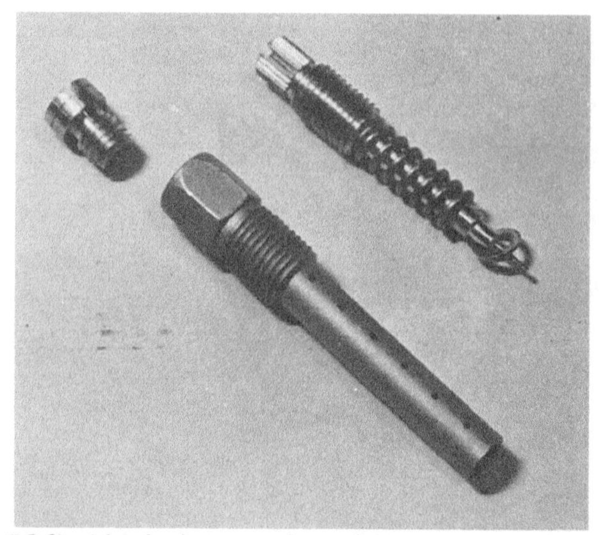
7.3 Check jets for damage or obstructions

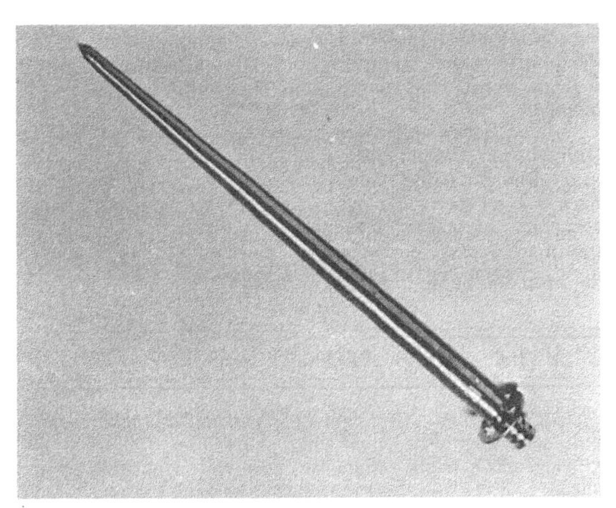

7.4 Needle should be straight and undamaged

jet assembly. It is pressed into position and thus must be cleaned in situ.
6 Examine the throttle valve for scoring or wear, renewing if badly damaged. If damage is evident, check the internal bore of the carburettor, and if necessary renew this also. Check that the needle is free from scoring or other damage and roll it on a flat surface to check that it has not become bent.

8 Carburettor: adjustments and settings

1 The various jet sizes, throttle valve cutaway and needle position are predetermined by the manufacturer and should not require modification. Check with the Specifications list at the beginning of this Chapter if there is any doubt about the types fitted.
2 Before any attempt at adjustment is made, it is important to understand which parts of the instrument control which part of its operating range. A carburettor must be capable of delivering the correct fuel/air ratio for any given engine speed and load. To this end, the throttle valve, or slide as it is often known, controls the volume of air passing through the choke or bore of the instrument. The fuel, on the other hand is regulated by the pilot and main jets, by the jet needle, and to some extent, by the amount of cutaway on the throttle valve.
3 As a rough guide, up to $\frac{1}{8}$ throttle is controlled by the pilot jet, $\frac{1}{8}$ to $\frac{1}{4}$ by the throttle valve cutaway, $\frac{1}{4}$ to $\frac{3}{4}$ throttle by the needle position and from $\frac{3}{4}$ to full throttle by the size of the main jet. These are only approximate divisions, which are by no means clear cut. There is a certain amount of overlap between the various stages.
4 If any particular carburation fault has been noted, it is a good idea to try to establish the most likely cause before dismantling or adjusting takes place. If, for example, the engine runs normally at road speeds, but refuses to tick over evenly, the fault probably lies with the pilot mixture system, and will most likely prove to be an obstructed jet. Whatever the problem may appear to be, it is worth checking that the jets are clear and that all the components are of the correct type. Having checked these points, refit the carburettor and check the settings as follows.
5 Set the pilot mixture screw to the position given in the Specifications Section. Start the engine, and allow it to attain its normal working temperature. This is best done by riding the machine for a few miles. Set the throttle stop screw to give a normal idling speed. Try turning the pilot mixture screw inwards by about $\frac{1}{4}$ turn at a time, noting its effect on the idling speed, then repeat the process, this time turning the screw outwards. The pilot mixture screw should be set in the position which gives the fastest consistent tickover. If desired, the tickover speed may be reduced further by lowering the throttle stop screw, but care should be taken that this does not cause the engine to falter and stop after the throttle twistgrip has been opened and closed a few times.

9 Carburettor: checking and setting the float height

1 It is important that the level of fuel in the float bowl is maintained at the prescribed height to avoid adverse effects on the mixture strength. It is worth noting that unless the correct float height is set, it will be impossible to set the remaining adjustments to obtain efficient running.
2 The float height is measured between the gasket face of the carburettor body and the bottom of the float. This should be done with the carburettor turned 90° to its normal position so that the weight of the float is not applied to the valve needle. The latter should just bear upon the valve seat when the measurement is made. The correct height should be 14.5 mm (0.57 in).
3 If adjustment is required it should be made by carefully bending the small tank which operates the valve needle. Note that a very small amount of movement at the tang will translate into a much larger change in the float height, so make any adjustments as fine as possible.

10 Air filter: general description and maintenance

1 The air used in combustion is drawn into the carburettor via an air filter element. This performs the vital job of removing dust and any other airborne impurities which would otherwise enter the engine, causing premature wear. It follows that the element must be kept clean and renewed, if damaged, as it will have an adverse effect on performance if neglected. Apart from the obvious problem of increased wear caused by a damaged element, a clogged or broken filter will upset the mixture setting, allowing it to become too rich or too weak.
2 The filter element is housed in a moulded plastic casing beneath the seat and is accessible via the left-hand side panel. The latter is a push-fit in three rubber bushes. The foam element is mounted on a metal frame which can be released by operating the spring catch at the rear of the assembly.
3 Follow the procedure given in Routine maintenance for cleaning and inspection of the element.

11 Exhaust system: general description and maintenance

1 The CB250 RS employs two separate one-piece exhaust systems. Each welded exhaust pipe and silencer assembly is retained at the exhaust port by split collets and a retaining flange, and at the silencer bracket by the rear footrest mounting. Frequent attention is unlikely to be necessary, apart from regular cleaning to protect the chromium plating. All mountings must be kept tight, especially at the exhaust ports.
2 Air leaks at the exhaust port, ie the joint between the exhaust pipe and the cylinder head, will cause mysterious backfiring when the machine is on overrun, as air will be drawn in causing residual gases to be ignited in the exhaust pipe. To this end, make sure that the composite sealing ring is renewed each time the system is removed.
3 There is no means of detaching the baffles from the silencer to facilitate their cleaning; usually the silencer will have reached the point of visible deterioration long before there is any chance of obstruction through carbon build-up.
4 The silencer, as its name implies, effectively reduces the exhaust noise to an acceptable level without having any adverse effects on engine performance. Do not tamper with or remove the baffles from within the silencer. Although a much louder exhaust note may give the impression of greater speed, this is rarely the case in practice. Tampering with the standard system, designed to give optimum performance with a low noise level,

will upset the balance and cause reduced performance, even though the changed exhaust note creates the illusion of speed.
5 If a non-standard exhaust system is chosen to replace the original fitment, do remember that this can invalidate any remaining warranty. Care must be taken when selecting exhaust systems. A good quality type will be as good or better than the original, and may allow the engine to develop more power, albeit at the expense of low to mid-range performance. Most exhaust changes will necessitate re-jetting of the carburettor, and this information will be available from any reputable manufacturer. Be wary of any system which does not come with re-jetting and performance details. It is best to avoid some of the cheapest systems. These may fail very quickly in use, can be illegally noisy and could even cause engine damage. For these reasons, it is best to chose a system of reputable manufacture.

12 Oil pump: removal, examination and reassembly

1 The oil pump will not normally require specific attention until a complete engine overhaul is necessary. If, however, it proves necessary to examine the pump for any reason, it can be removed from the crankcase after detaching the right-hand outer cover. Removal of the pump is covered specifically in Section 10 of Chapter 1. The various pump components should be checked for wear or damage as follows.
2 Examine each component for signs of scuffing and wear. Note especially the condition of the rotors and pump body. If these are at all worn, the pump must be renewed. Replace the spindle and rotors, and measure the clearance between the outer rotor and the pump body using feeler gauges. The nominal clearance is 0.15 – 0.18 mm (0.006 – 0.007 in). Renew the pump if the clearance is more than 0.25 mm ((0.010 in). Check the clearance between any two rotor peaks in a similar manner. The nominal clearance is 0.15 mm (0.006 in). Renew if worn to 0.2 mm (0.008 in).
3 The side float of the rotors can be checked by laying a straight edge across the pump face, and measuring the gap between it and the rotors. This gap should normally be 0.01 – 0.07 mm (0.0004 – 0.0028 in). If the measurement exceeds 0.12 mm (0.0047 in) the rotors and/or pump body should be renewed.
4 When reassembling the pump, lubricate each component with clean engine oil, making sure that it is worked around the rotors. The pump backplate is located by a dowel and is secured by a single screw. The pump body screws, when fitted, also secure the backplate. Note that to ensure correct alignment of the pump backplate, the screw should not be tightened fully until **after** the pump has been installed.

13 Oil filter: location and general description

1 The oil filtration arrangement is unusual by modern standards and relies on a single small gauze filter. This traps any particles which might otherwise be drawn up into the pump, causing damage to it and the moving parts of the engine and gearbox. It must be appreciated that any debris which is smaller than the holes in the gauze can pass freely into the lubrication system, and whilst this will not normally cause problems, any delay in changing the engine oil can lead to accelerated rates of engine wear. It follows that oil changes must be performed promptly at the specified intervals to prolong the life of the engine.
2 It is advisable to remove and clean the gauze screen at regular intervals, and as it is necessary to drain the oil content of the engine unit to do so, it is convenient to clean the screen at each oil change. To gain access to the filter screen it is necessary to remove the right-hand outer casing. The filter resides in a slot at the bottom of the casing and can be withdrawn using a pair of pointed-nose pliers. Clean the element by washing it thoroughly in clean petrol, making sure that all traces of contamination are removed. When clean and dry the element can be refitted and new engine oil added.

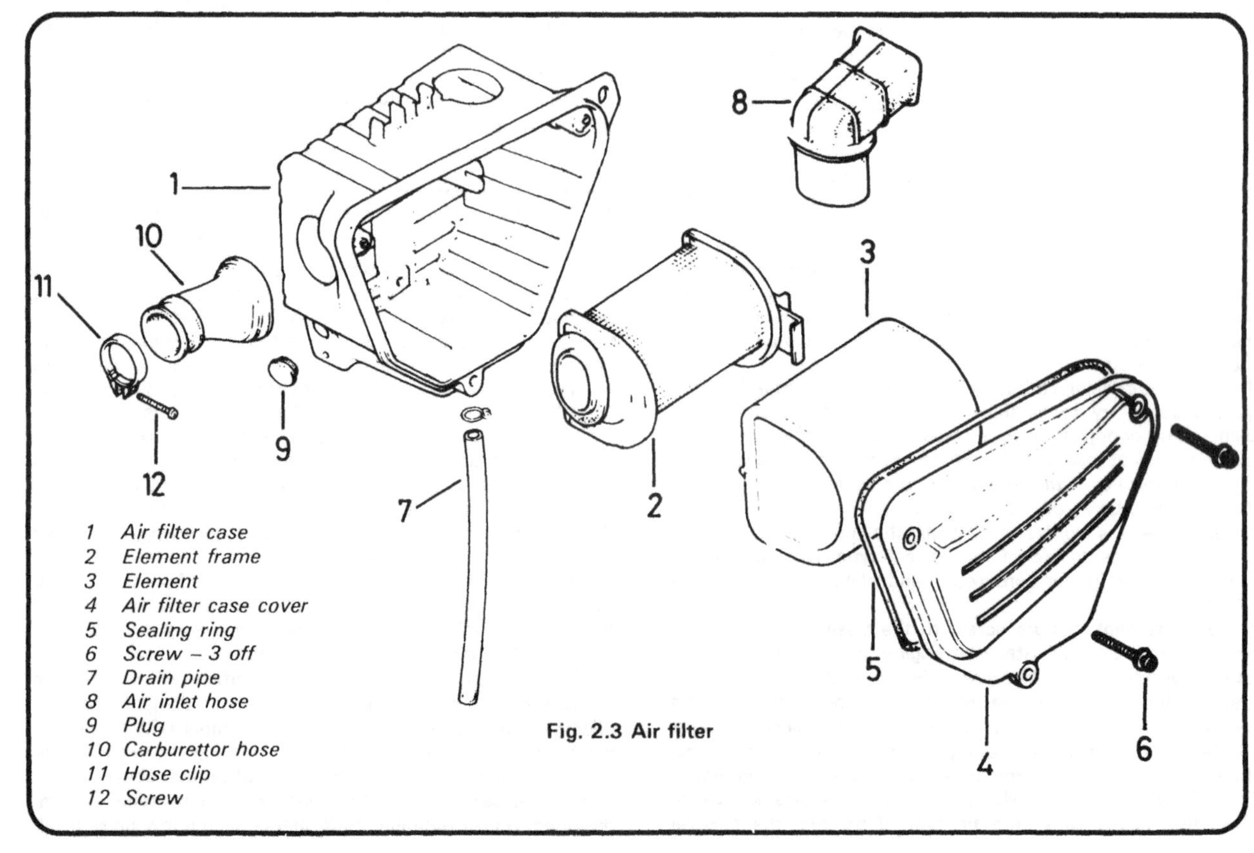

1 Air filter case
2 Element frame
3 Element
4 Air filter case cover
5 Sealing ring
6 Screw – 3 off
7 Drain pipe
8 Air inlet hose
9 Plug
10 Carburettor hose
11 Hose clip
12 Screw

Fig. 2.3 Air filter

12.2a Release single screw and remove pump cover

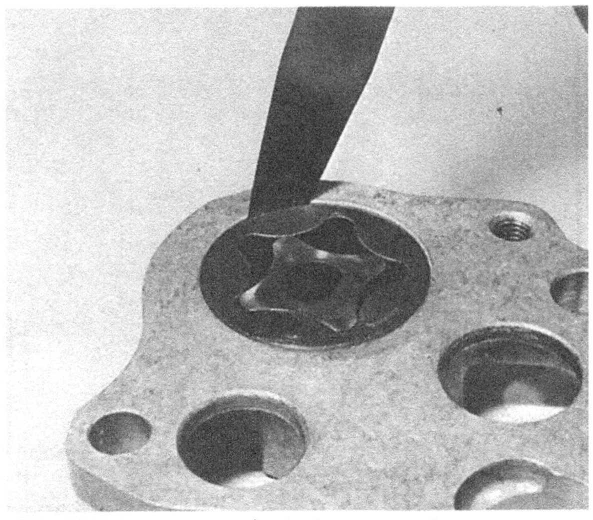

12.2b Measure outer rotor/body clearance as shown

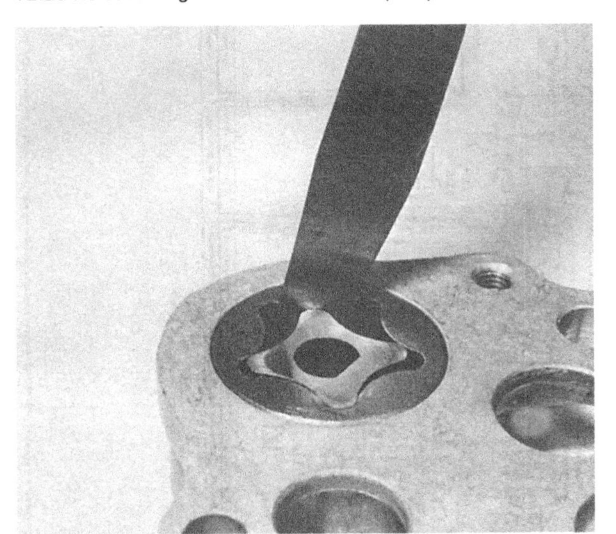

12.2c Check clearance between inner and outer rotors

12.3 Use straightedge to check rotor end float

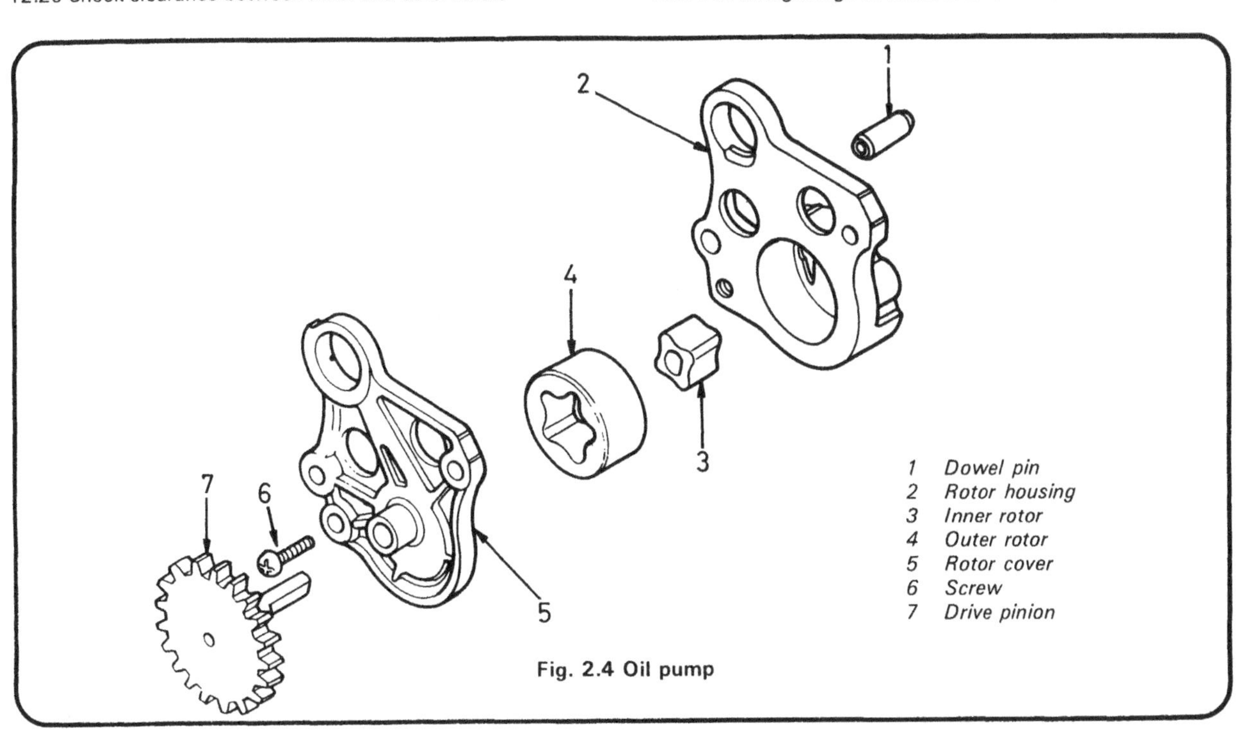

Fig. 2.4 Oil pump

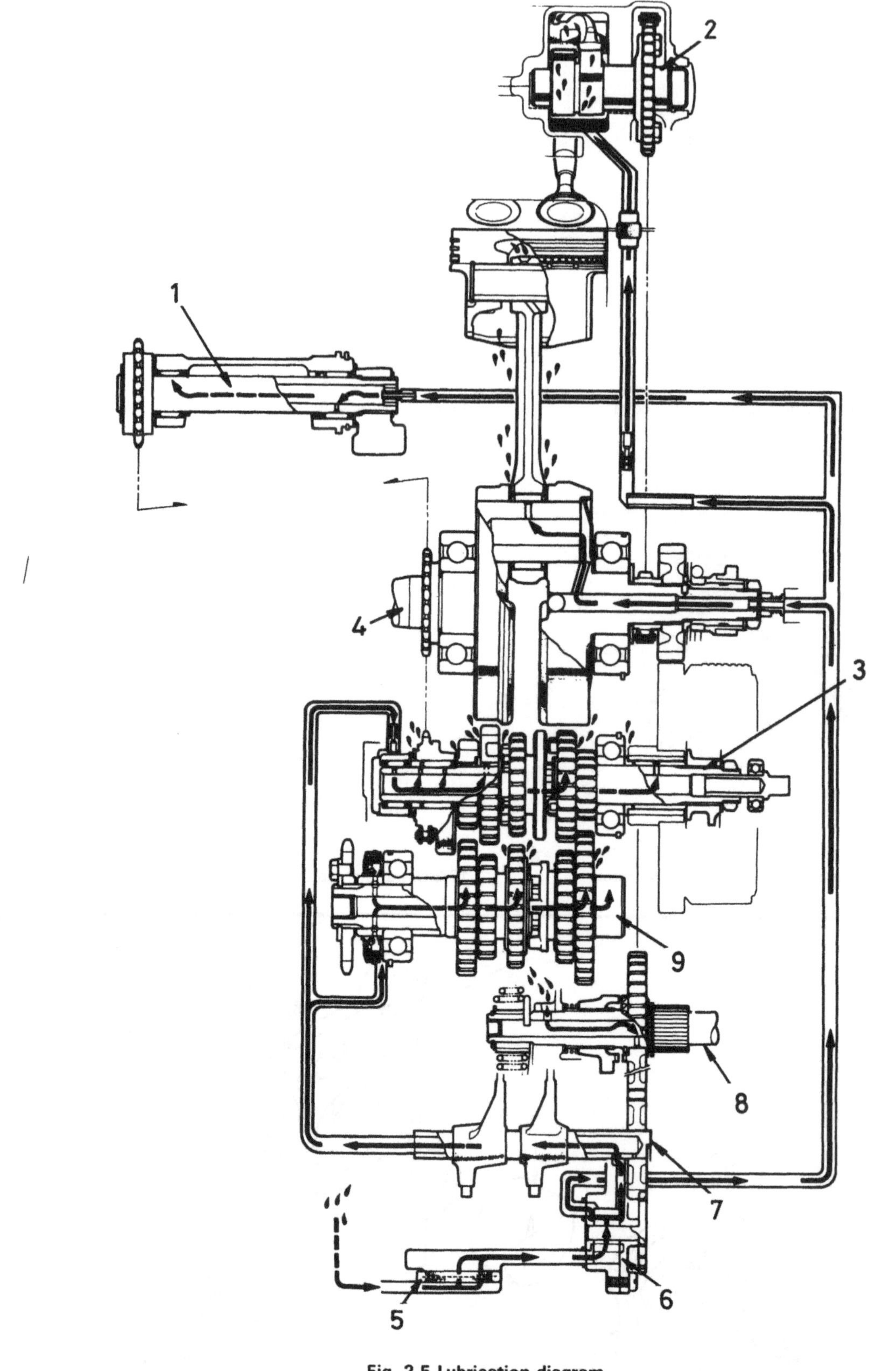

Fig. 2.5 Lubrication diagram

1	*Front balancer shaft*	*4*	*Crankshaft*	*7*	*Selector fork shaft*
2	*Camshaft*	*5*	*Oil filter screen*	*8*	*Kickstarter spindle*
3	*Mainshaft*	*6*	*Oil pump*	*9*	*Layshaft*

14 Fault diagnosis: fuel system and lubrication

Symptom	Cause	Remedy
Excessive fuel consumption	Air cleaner choked or restricted	Clean or renew.
	Fuel leaking from carburettor. Float sticking	Check all unions and gaskets. Float needle seat needs cleaning.
	Float damaged or leaking	Check and renew.
	Float level set too high	Check and adjust (see text)
	Badly worn or distorted carburettor	Replace.
	Jet needle setting too high	Adjust to figure given in Specifications.
	Main jet too large or loose	Fit correct jet or tighten if necessary
	Carburettor flooding	Check float valve and replace if worn.
Idling speed too high	Throttle stop screw in too far	
	Carburettor top loose	Adjust screw. Tighten top.
	Pilot screw incorrectly adjusted	Refer to relevant paragraph in this Chapter.
	Throttle cable sticking	Disconnect and lubricate or replace.
Engine dies after running for a short while	Blocked air hole in filler cap	Clean.
	Dirt or water in carburettor	Remove and clean out.
General lack of performance	Weak mixture; float needle stuck in seat.	Remove float chamber or float and clean.
	Air leak at carburettor joint	Check joint to eliminate leakage, and fit new O ring
Engine does not respond to throttle	Throttle cable sticking	See above.
	Petrol octane rating too low	Use higher grade (star rating) petrol.
Engine runs hot and is noisy	Lubrication failure	Stop engine immediately and investigate cause. Do not restart until cause is found and rectified.

Note: Incorrect ignition settings can give rise to symptoms similar to some of the above. Check the ignition system in conjunction with the fuel system to eliminate this possibility. (See Chapter 3 for details).

Chapter 3 Ignition system

For information relating to the CB250 RSD-C model, refer to Chapter 7

Contents

Specifications

Ignition system

Type	Capacitor discharge ignition (CDI) with mechanical advance

Sparking plug

Make	NGK or ND
Type	DR8ES or X27ESR-U
Gap	0.6 - 0.7 mm (0.024 - 0.028 in)

Ignition timing

Initial	12° BTDC @ 1200 rpm (F mark aligned)
Advance starts	13° BTDC @ 2250 rpm
Full advance	25° BTDC @ 3450 rpm

1 General description

The Honda CB250 RS is equipped with a CDI (capacitor discharge ignition) system. The system is powered by a source coil built into the alternator stator. Power from this coil is fed directly to the CDI unit mounted beneath the frame, where it passes through a diode which converts it to direct current (dc). The charge is stored in a capacitor at this stage.

The spark is triggered by the pulser assembly which is mounted on the right-hand end of the crankshaft. As the magnetic rotor passes the pulser coil a small alternating current (ac) pulse is induced. This enters the CDI unit where it is rectified by a second diode. The heart of the CDI unit is a component known as a thyristor. It acts as an electronic switch, which remains off until a small current is applied to its gate terminal. This causes the thyristor to become conductive, and it will remain in this state until any stored charge has discharged through the primary windings of the coil.

The sudden discharge of low-tension energy through the coil's primary windings in turn induces a high tension charge in the secondary coil. It is this which is applied to the centre electrode of the sparking plug, where it jumps the air gap to earth, igniting the fuel/air mixture.

As engine speed rises, it becomes necessary for the timing of the ignition spark to be advanced in relation to the crankshaft to allow sufficient time for the combustion of the air/fuel mixture to take place at the optimum position. This function is catered for by a conventional centrifugal automatic timing unit incorporated in the pulser rotor assembly.

2 CDI system: fault diagnosis

1 As no means of adjustment is available, any failure of the system can be traced to the failure of a system component or a simple wiring fault. Of the two possibilities, the latter is by far the most likely. In the event of failure, check the system in a logical fashion, as described below.

2 Remove the sparking plug, giving it a quick visual check, noting any obvious signs of flooding or oiling. Fit the plug into the plug cap and rest it on the cylinder head so that the metal body of the plug is in good contact with the cylinder head metal. The electrode end of the plug should be positioned so that sparking can be checked as the engine is spun over using the kickstart.

3 *Important note.* The energy levels in electronic systems can be very high. On no account should the ignition be switched on whilst the plug or plug cap are being held. Shocks from the HT circuit can be most unpleasant. Secondly, it is vital that the plug is in position and soundly earthed when the system is checked for sparking. The CDI unit can be seriously damaged if the HT circuit becomes isolated.

4 Having observed the above precautions, turn the ignition and engine kill switches to 'On' and kick the engine over. If the system is in good condition a regular, fat blue spark should be evident at the plug electrodes. If the spark appears thin or yellowish, or is non-existent, further investigation will be necessary. Before proceeding further, turn the ignition off and remove the key as a safety measure.

5 Ignition faults can be divided into two categories, namely

those where the ignition system has failed completely, and those which are due to a partial failure. The likely faults are listed below, starting with the most probable sources of failure. Work through the list systematically, referring to the subsequent sections for full details of the necessary checks and tests.

Total or partial ignition system failure

a) Loose, corroded or damaged wiring connections, broken or shorted wiring between any of the component parts of the ignition system
b) Faulty main switch or engine kill switch
c) Faulty ignition coil
d) Faulty CDI unit
e) Faulty alternator
f) Faulty pulser assembly

3 CDI system: checking the wiring

1 The wiring should be checked visually, noting any signs of corrosion around the various terminals and connectors. If the fault has developed in wet conditions it follows that water may have entered any of the connectors or switches, causing a short circuit. A temporary cure can be effected by spraying the relevant area with one of the proprietary de-watering aerosols, such as WD40 or similar. A more permanent solution is to dismantle the switch or connector and coat the exposed parts with silicone grease to prevent the ingress of water. The exposed backs of connectors can be sealed off using a silicone rubber sealant.

2 Light corrosion can normally be cured by scraping or sanding the affected area, though in serious cases it may prove necessary to renew the switch or connector affected. Check the wiring for chafing or breakage, particularly where it passes close to part of the frame or its fittings. As a temporary measure, damaged insulation can be repaired with PVC tape, but the wire concerned should be renewed at the earliest opportunity.

3 Using the wiring diagram at the end of the manual, check each wire for breakage or short circuits using a multimeter set on the resistance scale or a dry battery and bulb wired as shown in the accompanying illustration. In each case, there should be continuity between the ends of each wire.

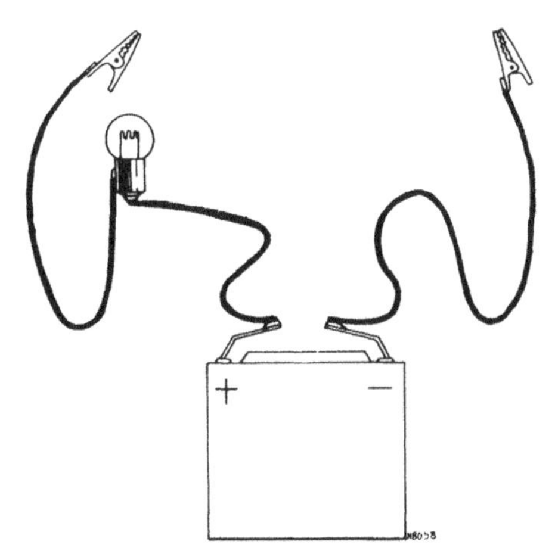

Fig. 3.1 Continuity testing circuit

4 CDI system: checking the ignition and engine kill switches

1 The ignition system is controlled by the ignition switch or main switch, which is housed at the centre of the instrument console, and by the engine kill switch incorporated in the right-hand handlebar switch cluster. The ignition switch has six terminals and leads, of which two are involved in controlling the ignition system. These are the IG terminal (Black/white lead) and the E terminal (Green lead). The two terminals are connected when the switch is in the 'Off' position and prevent the ignition system from functioning by shorting the CDI unit to earth. A duplicate set of terminals with identical wiring colours form the kill switch, thus when either switch is set to 'Off' the ignition switch is rendered inoperative. When both switches are set to the 'On' position, the CDI to earth connection is broken, and the system is allowed to function.

2 If the operation of either switch is suspect reference should be made to the wiring diagram at the end of this book. The switch connections are shown in diagrammatic form and indicate which terminals are connected in the various switch positions. The wiring from the switches can be traced back to their respective connectors where test connections can be made most conveniently.

3 The purpose of the test is to check whether the switch connections are being made and broken as indicated by the above diagrams. In the interests of safety the test is made with the machine's battery disconnected, thus avoiding accidental damage to the CDI system or the owner. The test can be made with a multimeter set on the resistance scale, or with a simple dry battery and bulb arrangement, as shown in the accompanying line drawing. Connect one probe lead to each terminal and note the reading or bulb indication in each switch position.

4 If the test indicates that the black/white lead is earthed irrespective of the switch position, check that the engine kill switch is set at the 'Run' position. If this fails to affect the result, trace and disconnect the ignition (black/white) and earth (green) leads from the ignition switch. Repeat the test with the switch isolated. If no change is apparent, the switch should be considered faulty and renewed.

5 If the ignition switch works normally when isolated, the fault must lie in the black/white lead between the CDI unit and the ignition and kill switches or in the kill switch itself. As already mentioned, the ignition and kill switches perform the same function, each earthing the ignition circuit when set on the 'Off' position, thus unless both are disconnected from earth, the ignition circuit will remain inoperative.

6 The kill switch is checked in the same manner as described above. If a fault is discovered in either switch, try cleaning it with a water dispersant aerosol spray, such as WD40, Contect or similar. If this fails to effect a cure, it may prove necessary to renew the switch, a decision not to be taken lightly in view of the cost. One solution would be to try to obtain a good secondhand switch from a motorcycle breaker.

7 The ignition switch is mounted on the underside of the instrument console, and may be removed after the latter has been detached. The console is retained by two bolts, as is the switch itself. The kill switch is housed in the right-hand handlebar switch cluster and is removed by separating the two halves of the switch. Although repair should be considered impracticable, it may prove to be worthwhile attempting it if the switch is otherwise useless. The ignition switch, however, is a sealed unit and cannot be dismantled.

5 Ignition coil: location and testing

1 The ignition coil is a sealed unit, and will normally give long service without need for attention. It is mounted beneath the frame gusseting to the rear of the steering head and is covered in use by the fuel tank. It follows that it will be necessary to remove the tank in order to gain access to the coil.

2 If a weak spark and difficult starting causes the performance of the coil to be suspect, it should, in general, be tested by a Honda service agent or an auto-electrical expert. They will have the necessary appropriate test equipment. It is, however, possible to perform a number of basic tests, using a multimeter with ohms and kilo ohms scales. The primary winding resistance should be checked by connecting one of the

meter probe leads to the Lucar terminal and the other earthed against the coil mounting lug. The secondary windings are checked by connecting the probe leads to the high tension lead, having removed the plug cap, and to the coil mounting lug.

Primary winding resistance 0.2 – 0.8 ohms
Secondary winding resistance 8 – 15 kilo ohms

3 Should any of these checks not produce the expected result, the coil should then be taken to a Honda service agent or auto-electrician for a more thorough check. If the coil is found to be faulty, it must be replaced; it is not possible to effect a satisfactory repair.

6 CDI unit: location and testing

1 The CDI unit takes the form of a sealed metal box mounted beneath the fuel tank. In the event of malfunction the unit may be tested in situ after the fuel tank has been removed and the wiring connectors traced and separated. Honda advise against the use of any test meter other than the Sanwa Electric Tester (Honda part number 07308-0020000) or the Kowa Electric Tester (TH-5H), because they feel that the use of other devices may result in inaccurate readings.

2 Most owners will find that they either do not possess a multimeter, in which case they will probably prefer to have the unit checked by a Honda Service Agent, or own a meter which is not of the specified make or model. In the latter case, a good indication of the unit's condition can be gleaned in spite of inaccuracies in the readings. If necessary, the CDI unit can be taken to a Honda Service Agent or auto-electrical specialist for confirmation of its condition.

3 The test details are given in the accompanying illustration in the form of a table of meter probe connections with the expected reading in each instance. If an ordinary multimeter is used the resistance range may be determined by trial and error. The diagram illustrates the CDI unit connections referred to in the table. For owners not possessing a test meter the unit or the complete machine can be taken to a Honda Service Agent for testing.

4.7 Ignition switch is bolted to the underside of instrument panel

5.1 Ignition coil is mounted beneath the fuel tank

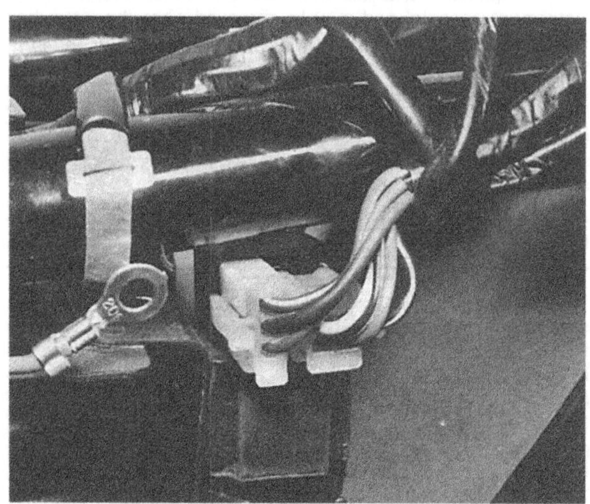

6.1 CDI unit is mounted forward of the battery compartment

6.3 CDI unit showing terminals referred to in tests

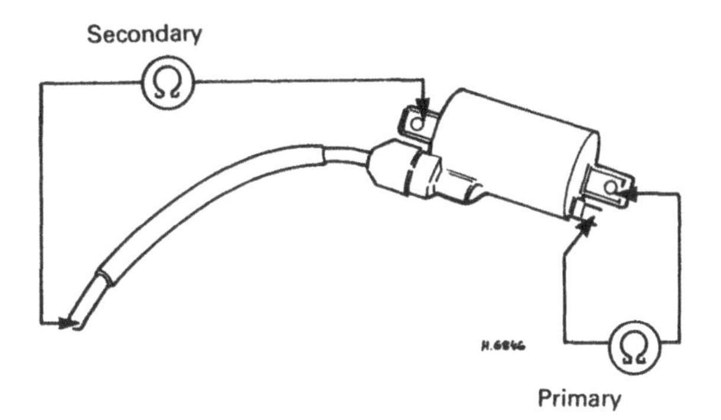

Fig. 3.2 Ignition coil resistance check

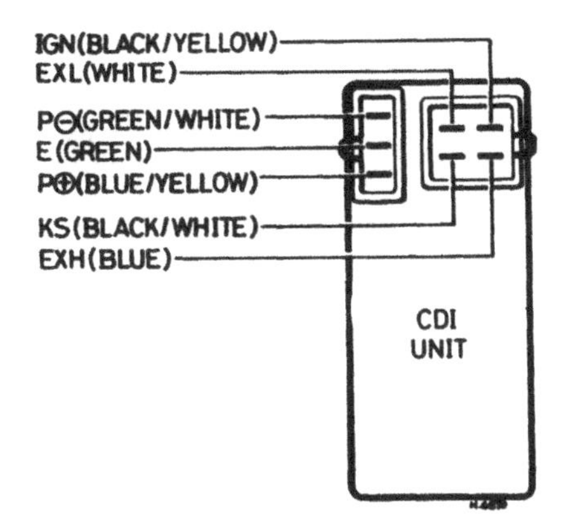

SANWA: x kΩ
KOWA: x 100Ω

PROBE ⊖ \ ⊕ PROBE	KS	EXH	EXL	E	P⊕	P⊖	IGN
KS		∞	∞	∞	∞	∞	∞
EXH	5~100		∞	∞	∞	∞	∞
EXL	∞	∞		∞	∞	∞	∞
E	∞	∞	1~50		∞	0	∞
P⊕	∞	∞	2~60	2~60		2~60	∞
P⊖	∞	∞	1~50	0	∞		∞
IGN	∞	∞	∞	∞	∞	∞	

Fig. 3.3 CDI unit testing

7 Alternator source coil: testing

1 The ignition system is powered by a source coil which is built into the alternator stator. It follows that if the alternator malfunctions it will affect the operation of the ignition system, possibly without affecting the entire electrical system. Six leads exit from the stator in two groups of three, terminating in two three-pin connectors. One of these carries the three yellow main output leads from the alternator stator to regulator/rectifier unit. The second connector carries the white and blue leads from the ignition source coil to the CDI unit, plus the light green/red lead from the neutral switch.

2 The stator can be checked by measuring the resistance at the two connectors using a multimeter set on the ohms scale. The values should be as shown below.

Yellow to yellow leads	*0.3 - 0.53 ohms*
White to earth	*259 - 351 ohms*
White to blue	*73 - 99 ohms*

If the readings fall outside these limits the stator should be checked by a Honda Service Agent and renewed where necessary.

8 Pulser assembly: testing

1 The pulser assembly consists of a small coil unit, mounted on the inside of the right-hand outer casing, and a magnetic rotor (reluctor) which is attached to the end of the crankshaft. As the crankshaft rotates, the rotor tip passes the coil pole inducing a small trigger current which is then fed to the thyristor gate in the CDI unit. It will be appreciated that, in a four-stroke application, this system produces two sparks per complete engine cycle, one of which occurs during the exhaust stroke and is thus wasted. This arrangement is intentional, having no effect on the running of the engine, but making the triggering of the spark easier to contrive. Neither the stator nor rotor parts of the unit are adjustable, making timing adjustment impossible and thus regular timing checks unnecessary.

2 In the event that the pulser assembly fails to operate, it will be seen that the CDI unit will not be triggered, and thus sparks at the plug electrodes will be noticeably lacking. Generally speaking, the assembly is not likely to cause problems. It is theoretically possible for the rotor to become demagnetised to the point where sparking becomes unreliable, but in practice the tiny current required to trigger the CDI unit is invariably produced. A short or open circuit in the pulser coil is more likely, and can be checked by measuring the coil resistance.

3 Trace the pulser leads back to the connector beneath the fuel tank. Separate the connector and measure the resistance between the green and the blue/yellow wires. In good condition, the pulser coil should show a resistance of 90 - 100 ohms. A reading of infinity or zero resistance will be indicative of an open or short circuit respectively, and will necessitate renewal of the coil. Should renewal be required, take the rotor assembly to ensure that the new coil unit has the same matching mark. An unmatched pulser coil and rotor can result in incorrect ignition timing

8.1 Pulser coil is mounted on inside of outer cover

9 Ignition timing: checking

1 As a result of the adoption of electronic ignition, periodic checking of the ignition timing should not be necessary and, indeed, no provision is given for adjustment. Checking the timing is worthwhile, however, if a new pulser or reluctor is fitted, and it is also a useful means of checking the function of the mechanical ATU if its performance is suspect.
2 The ignition timing can be checked only whilst the engine is running using a stroboscopic lamp and thus a suitable timing lamp will be required. The inexpensive neon lamps should be adequate in theory, but in practice may produce a pulse of such low intensity that the timing mark remains indistinct. If possible, one of the more precise xenon tube lamps should be employed powered by an **external** source of the appropriate voltage.
3 The ignition timing is checked by reference to timing marks scribed on the periphery of the generator rotor and a fixed index mark on the generator cover. To gain access to the marks remove the upper inspection cap from the cover. To check the ignition timing connect the stroboscope to the machine's ignition system high-tension or low-tension circuit as directed by the lamp's manufacturer. Start the engine and aim the lamp at the generator rotor. If the ignition timing is correct the 'F' mark will align with the index mark at 1200 rpm. Raise the speed of the engine until the full advance speed of 3450 rpm is obtained. At this speed the index mark should align with the two parallel advance marks scribed on the generator rotor. The transition in the timing from retarded to full advance should take place smoothly in proportion to the speed. If it can be seen that the movement is erratic or if full advance cannot be reached then there is some indication that the advance mechanism is not functioning correctly. Refer to the following Section for details of ATU inspection.

10 Automatic timing unit (ATU): location, function and testing

1 Despite the adoption of the CDI system in place of the earlier coil-and-contact breaker systems, Honda have elected to retain a mechanical centrifugal advance system in the form of an automatic timing unit. This device is necessary to cause the ignition spark to take place progressively earlier as the engine speed increases, so that the fuel/air mixture is given time to burn in a manner which will produce the maximum amount of useful work.
2 The automatic timing unit consists of a baseplate fitted with a central spindle, by which it is retained to the crankshaft end through a taper and securing bolt, and two pivot pins arranged at the outer edge of the plate at 180° to each other. Two bobweights are fitted over the pivot pins and anchored by a pair of small tension springs. Tangs on the weights engage in the ignition pulser rotor (reluctor) which is mounted concentrically over the centre spindle.
3 As the engine runs, the ATU is spun on the end of the camshaft. This results in the weights being flung outwards by centrifugal force, and in doing so, they cause the rotor to move in relation to the camshaft. The weights are controlled by the light tension springs, and these pull the weights back as the engine slows down.
4 To gain access to the ATU, it is necessary to remove the right-hand outer cover after draining the engine oil. The rotor and ATU unit is secured by a single nut. It may prove necessary to lock the crankshaft whilst the nut is removed. This can best be accomplished by selecting top gear and applying the rear brake. Remove the nut and slide the ATU off the crankshaft end. It is not necessary to disturb the spring-loaded oil feed quill.
5 Examine each component for wear, checking that none of the anchor and pivot pins have become loose in the backplate. It may be possible to secure loose pins by re-riveting them, but often the whole assembly will require renewal. Similarly, only a small amount of wear can be tolerated before sloppy operation allows the ignition setting to wander.
6 There is little point in attempting to dismantle the unit since the component parts are not available separately. If a new unit is required it is essential that it matches the pickup or pulser coil and that the correct unit for any given model year is fitted. To this end, the engine and frame number should be given in full.
7 The action of the ATU during operation can be checked visually using a stroboscopic lamp as discussed in the preceding Section.

10.1 ATU and rotor assembly should be checked for wear

11 Sparking plug: checking and setting the gap

1 The Honda CB250 RS is fitted as standard with either a NGK DR8ES or Nippondenso (ND) X27ESR-U sparking plug. In most operating conditions the standard plug should prove satisfactory. If the machine is used continually in temperatures below 5°C (41°F) an NGK DR8ES-L or ND X24ESR-U plug may be fitted.
2 The correct electrode gap is 0.6 - 0.7 mm (0.024 - 0.028 in). The gap can be assessed using feeler gauges. If necessary, alter the gap by removing the outer electrode, preferably using a proper electrode tool. **Never** bend the centre electrode, otherwise the porcelain insulator will crack, and may cause damage to the engine if particles break away whilst the engine is running.
3 After some experience the sparking plug electrodes can be used as a reliable guide to engine operating conditions. See accompanying photographs.
4 It is advisable to carry a new spare sparking plug on the machine, having first set the electrodes to the correct gap. Whilst sparking plugs do not fail often, a new replacement is well worth having if a breakdown does occur.
5 Never overtighten a sparking plug otherwise there is risk of stripping the threads from the cylinder head, especially as it is cast in light alloy. A stripped thread can be repaired without having to scrap the cylinder head by using a 'Helicoil' thread insert. This is a low-cost service, operated by a number of dealers.
6 Before replacing a sparking plug into the cylinder head coat the threads sparingly with a graphited grease to aid future removal. Use the correct size spanner when tightening the plug otherwise the spanner may slip and damage the ceramic insulator. The plug should be tightened sufficiently to seat firmly on the sealing washer, and no more.

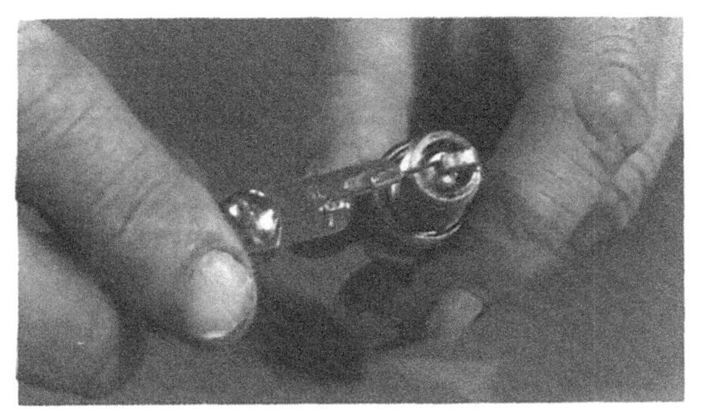

Spark plug maintenance: Checking plug gap with feeler gauges

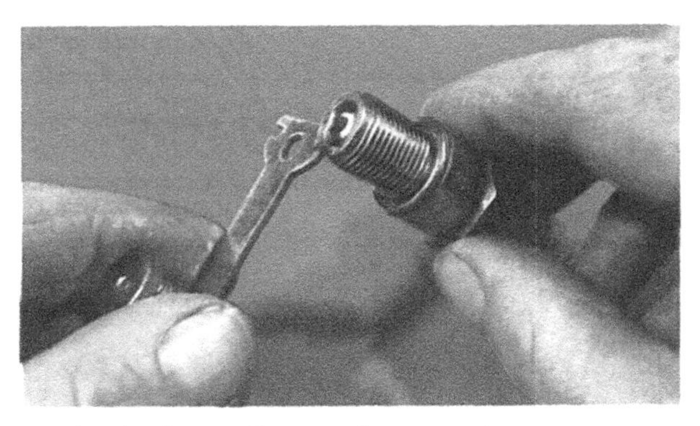

Altering the plug gap. Note use of correct tool

Spark plug conditions: A brown, tan or grey firing end is indicative of correct engine running conditions and the selection of the appropriate heat rating plug

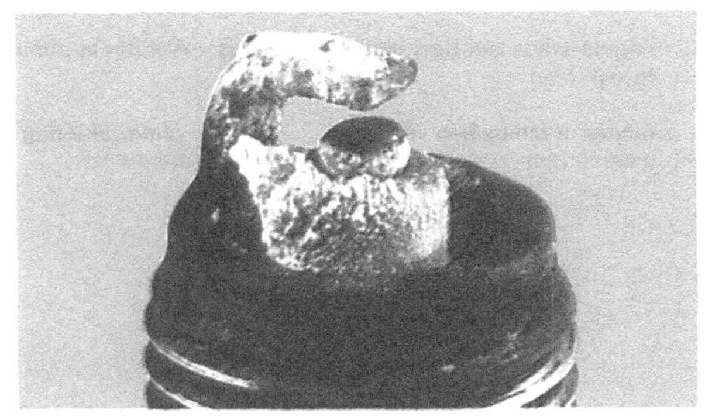

White deposits have accumulated from excessive amounts of oil in the combustion chamber or through the use of low quality oil. Remove deposits or a hot spot may form

Black sooty deposits indicate an over-rich fuel/air mixture, or a malfunctioning ignition system. If no improvement is obtained, try one grade hotter plug

Wet, oily carbon deposits form an electrical leakage path along the insulator nose, resulting in a misfire. The cause may be a badly worn engine or a malfunctioning ignition system

A blistered white insulator or melted electrode indicates over-advanced ignition timing or a malfunctioning cooling system. If correction does not prove effective, try a colder grade plug

A worn spark plug not only wastes fuel but also overloads the whole ignition system because the increased gap requires higher voltage to initiate the spark. This condition can also affect air pollution

12 Fault diagnosis: ignition system

Symptom	Cause	Remedy
Engine will not start	Spark at plug weak or non-existent	Fouled or faulty plug - renew Check ignition wiring and connections Plug cap, HT lead or coil faulty - renew CDI unit or pulser faulty - renew
Engine starts but runs erratically	Intermittent or weak spark	Check system as described above
As above, but engine tends to kick back when started. Engine runs hot	Automatic timing unit jammed open	Examine and clean or renew as required
Engine starts but runs sluggishly. Tends to overheat	Automatic timing unit jammed closed	See above
Ignition system fails when used in rain	Water shorting ignition or engine kill switch	Use water displacing spray such as WD40 or Contect to locate fault and effect temporary cure. Waterproof component at a later date

Chapter 4 Frame and forks

For information relating to the CB250 RSD-C model, refer to Chapter 7

Contents

Specifications

Frame ..... Welded steel, engine used as stressed member

Forks

Type	Oil damped telescopic
Oil capacity	158 ± 2.5 cc per leg
Oil grade	ATF (automatic transmission fluid) or fork oil
Fork spring free length	495.9 mm (19.5 in)
Service limit	477 mm (18.8 in)

Rear suspension

Type	Swinging arm
Suspension units	Gas filled, coil spring, hydraulically damped
Spring free length	199.5 mm (7.85 in)
Service limit	193.5 mm (7.62 in)
Swinging arm pivot clearance	0.2 - 0.3 mm (0.008 - 0.012 in)
Service limit	0.8 mm (0.032 in)

Torque wrench settings	**kgf m**	**lbf ft**
Fork crown nut	9.0 - 12.0	65 - 87
Upper yoke pinch bolts	0.9 - 1.3	7 - 9
Lower yoke pinch bolts	1.8 - 2.5	13 - 18
Fork spindle clamp nuts	1.8 - 2.5	13 - 18
Front wheel spindle nut	5.0 - 8.0	36 - 58
Swinging arm pivot nut	6.0 - 8.0	43 - 58
Rear suspension mountings	3.0 - 4.0	22 - 29
Rear wheel spindle nut	8.0 - 10.0	60 - 72
Brake torque arm nuts	1.8 - 2.5	13 - 18

1 General description

The Honda CB250 RS utilises a welded tubular steel frame built around a fabricated sheet steel main spine. The spine section is built up around the steering head and includes the top tube and part of the front down tube. No frame cradle is employed, the engine unit acting as a stressed member of the chassis.

Front suspension is by means of conventional oil-damped telescopic forks, whilst the rear suspension is provided by a pivoted rear fork controlled by oil-damped coil spring suspension units.

2 Front fork removal: general

1 It is unlikely that the forks will require removal from the frame unless the fork seals are leaking or accident damage has been sustained. In the event that the latter has occurred, it should be noted that the frame may also have become bent, and whilst this may not be obvious when checked visually, could prove to be potentially dangerous.
2 If attention to the fork legs only is required, it is unnecessary to detach the complete assembly, the legs being easily removed individually.
3 If attention to the steering head assembly is required it is possible to remove the lower yoke with the fork legs still in place, if desired. It should be noted, however, that this procedure is hampered by the unwieldy nature of the assembly, and it is recommended that the fork legs be removed prior to dismantling the steering head and fork yokes.
4 Before dismantling work can begin it will be necessary to arrange the machine so that the front wheel is raised clear of the ground. This is best done by lashing the rear of the machine down, either to a fixed object in the workshop or to a suitable weight.
5 Slacken and remove the single screw which retains the speedometer drive cable to the front wheel gearbox, refitting the screw to avoid its loss. Straighten and remove the split pin which locates the castellated front wheel spindle nut. Slacken and remove the nut, then release the clamp assembly at the right-hand end of the spindle. The spindle can now be withdrawn to clear the left-hand lower leg, and the wheel lowered clear of the forks.
6 Before the fork legs can be removed it will be necessary to remove the front mudguard and the front brake caliper. The mudguard is retained by two bolts screwed into each lower leg. Remove the two caliper mounting bolts and lift the caliper clear of the fork. Do not allow it to hang from the hydraulic hose because it is easily damaged. Tie the caliper to the frame to prevent damage and as a precaution slip a wooden wedge between the brake pads in case the brake lever is inadvertently operated during dismantling. This will prevent the ejection of the caliper piston.

3 Front forks: removing the fork legs from the yokes

1 It is not necessary to remove the complete headstock assembly if attention to the fork legs alone is required. The instructions in Section 2 of this Chapter should be followed, then proceed as described below.
2 Slacken the upper and lower pinch bolts which retain each fork leg. It should now be possible to pull and twist the fork legs downward to disengage them from the yokes. If necessary, a wooden drift can be used to knock the legs downward and clear of the yokes.

2.5a Release the single screw which secures the speedometer cable

2.5b Remove split pin and nut, and withdraw spindle to release wheel

2.6 Mudguard is bolted to inside of lower leg

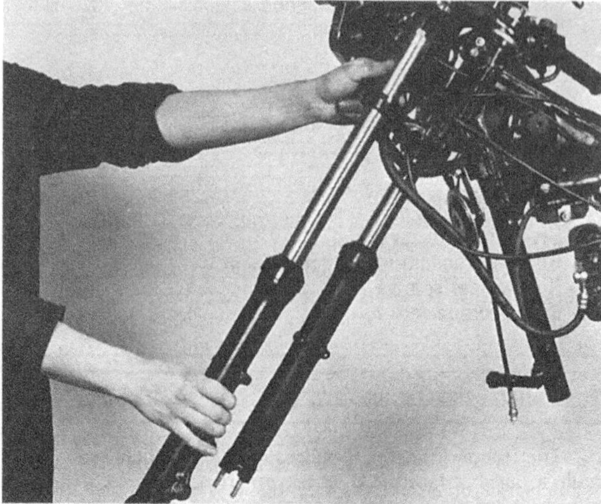
3.2 Slacken pinch bolts and free fork legs by pulling downwards

4 Steering head assembly: dismantling

1 Before the steering head itself can be removed it will be necessary to dismantle the front wheel and fork legs as described in Sections 2 and 3. To avoid damage and to provide access the fuel tank must be removed. Slacken and remove the two bolts which secure the seat to the frame, lifting the seat clear to reveal the tank mounting bolt. Turn the fuel tap to 'Off' and prise off the fuel pipe. Remove the tank mounting bolt to free the rear of the tank then lift and pull the tank rearwards to disengage the mounting rubbers at the front. Place the tank and seat to one side to await reassembly.
2 Release the two headlamp securing screws from the underside of the rim and lift the lens and reflector assembly clear of the shell. Disconnect the headlamp and parking lamp leads and place the unit to one side. Free the wiring from the headlamp shell by separating the connectors, noting that the wires are colour-coded and thus pose no problem when the time comes for reassembly. Push the wiring and connectors out through the apertures at the rear of the shell. The shell can now be removed after its two retaining bolts have been released.
3 Slacken the knurled rings which retain the speedometer and tachometer drive cables to their respective instrument heads and pull the cables clear of the steering head area, then trace and disconnect the leads to the instrument panel and ignition switch. Remove the two rubber-mounted bolts which retain the instrument panel and place it to one side. Remove the plastic cover and cable guide from the lower yoke.
4 It is now necessary to free the handlebar assembly from the top yoke to permit its removal. The official method is to remove the brake master cylinder, switches and clutch lever before releasing the handlebar, but it is possible to leave the assembly intact, removing the yokes after the handlebar clamps have been removed and the handlebar assembly pulled clear. The handlebar is retained by two clamps, each of which is secured by two Allen screws concealed below plastic caps. As the handlebar is lifted clear of the top yoke disengage and remove the headlamp sub-frame together with the front direction indicator lamps.
5 Remove the steering stem top nut, taking care not to damage the chromium plating. The top yoke can now be removed by pulling it upwards and clear of the steering stem. If necessary, tap the underside of the yoke to free it, using a wooden block to prevent damage to the yoke or its paint finish.
6 The lower yoke and steering stem can be removed after the slotted adjuster nut has been unscrewed using a C-spanner. Note that the steering head bearings are of the uncaged ball type, and some provision must be made to catch these as they drop free. Start by placing a large sheet of cloth or an old blanket beneath the steering head. This will catch any balls which drop, preventing them from bouncing off in all directions. As the nut is slackened place one hand below the lower race to catch the balls and remove any which stick to the cup or cone. The balls in the upper race will usually stay in position, but should be removed after the lower yoke has been freed. There should be a total of 37 $\frac{1}{4}$ in steel balls; 19 from the lower races and 18 from the upper race.
7 Check the bearing balls and races as described in Section 5, then reassemble the steering head by reversing the dismantling order. Place the balls in position, holding them in place with grease whilst the lower yoke is refitted. Fit the slotted adjuster nut finger-tight, then use a C-spanner to tighten the nut until a firm resistance is felt. The nut should then be backed off by $\frac{1}{8}$ turn. The object is to remove all traces of free play without placing any preload on the bearings. Note that overtight head races will make the steering stiff and unresponsive and may destroy the bearings. Loose bearings are equally undesirable and will have a serious affect on handling.
8 Place the top yoke in position but do not tighten the domed top nut at this stage. The fork legs should be temporarily refitted to align the yokes whilst the top nut is tightened to 9.0 – 12.0 kgf m (65 - 87 lbf ft). The fork legs can now be withdrawn again whilst assembly continues.
9 Note that the handlebar has a punch mark which should be aligned with the horizontal face of the split clamps. The clamp halves themselves have punch marks which should be positioned forward of the handlebar. Tighten the front clamp screws fully, then tighten the rear clamp screws. Where practicable, tighten the screws to 1.8 - 3.0 kgf m (13 - 22 lbf ft). Further punch marks are provided on the outer ends of the handlebar as an aid to aligning the switches and master cylinder.
10 Assemble the headlamp sub-frame and the instrument panel. Fit the headlamp shell and align the punch marks on it and its bracket when tightening the mounting bolts. Pass the wiring connectors into the shell and reconnect the wiring, noting the colour coding. Fit the headlamp unit and reconnect the speedometer and tachometer drive cables, followed by the plastic cover on the lower yoke. Complete assembly by installing the front forks and wheel as described in Section 11.

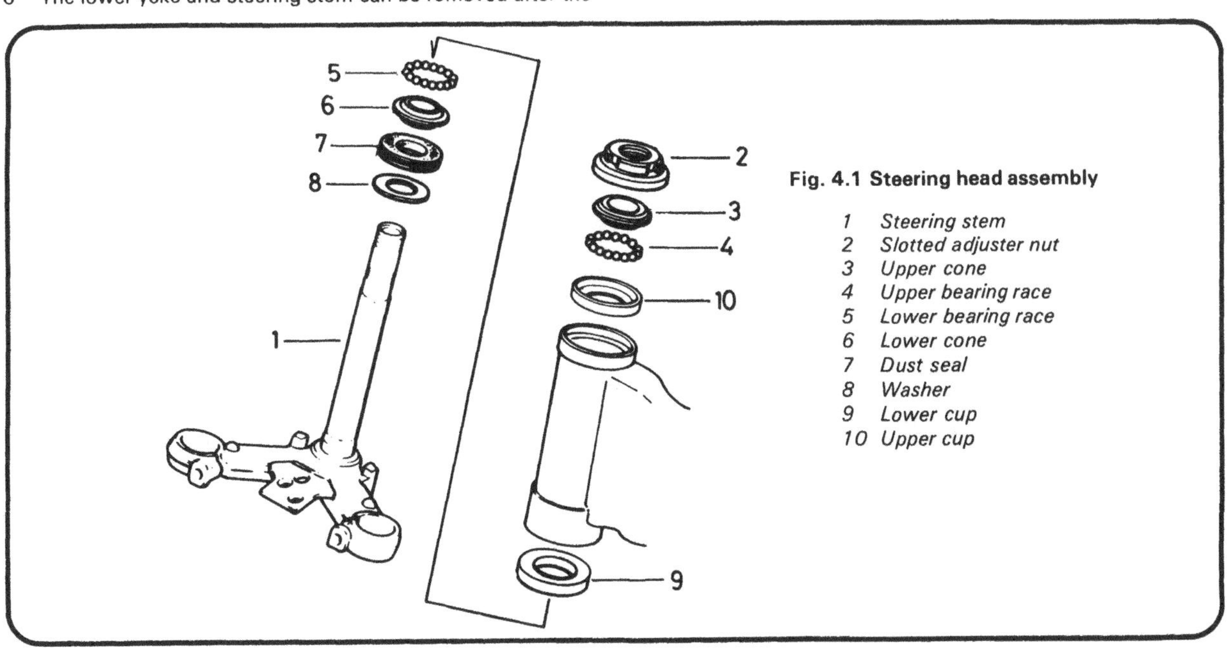

Fig. 4.1 Steering head assembly

1 *Steering stem*
2 *Slotted adjuster nut*
3 *Upper cone*
4 *Upper bearing race*
5 *Lower bearing race*
6 *Lower cone*
7 *Dust seal*
8 *Washer*
9 *Lower cup*
10 *Upper cup*

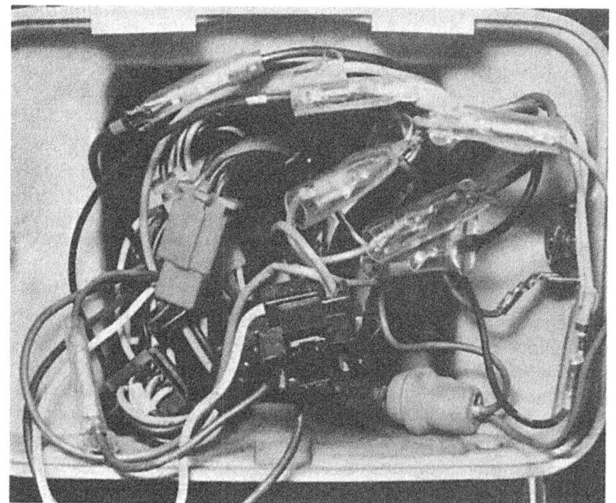
4.2 Disconnect wiring and remove headlamp shell

4.3a Detach speedometer and tachometer cables

4.3b Instrument panel is secured by two bolts (arrowed)

4.4 Remove caps and slacken clamp screws

4.9 Note alignment marks when refitting handlebar clamps

5 Steering head bearings: examination and renovation

1 Before commencing reassembly of the forks examine the steering head races. They are unlikely to wear out under normal circumstances until a high mileage has been covered. If, however, the steering head bearings have been maladjusted, wear will be accelerated. If, before dismantling, the forks had had a pronounced tendency to stick in one position when turned, often in the straight ahead position, the cups and cones are probably indented and need to be renewed.
2 Examine the cups and cones carefully; it is not necessary to remove the cups for this. The bearing tracks should be polished and free from indentations, cracks or pitting. If signs of wear are evident, the cups and cones must be renewed. In order for the straight line steering on any motorcycle to be consistently good, the steering head bearings must be absolutely perfect. Even the smallest amount of wear on the cups and cones may cause steering wobble at high speeds and judder during heavy front wheel braking. The cups and cones are an interference fit on their respective seatings and can be tapped from position using a suitable long drift.
3 Examine the ball bearings. Ball bearings are relatively cheap. If the originals are marked or discoloured they must be renewed as a complete set. To hold the steel balls in place

during reassembly of the fork yokes, pack the bearings with grease. The upper race is fitted with eighteen $\frac{1}{4}$ inch (No 8) steel balls and the lower race, nineteen balls of the same size. Although each race has room for an extra steel ball it must not be fitted. The small gap allows the bearings to work correctly, stopping them skidding on each other and accelerating the rate of wear.

6 Fork yokes: examination

1 To check the top yoke for accident damage, push the fork stanchions through the bottom yoke and fit the top yoke. If it lines up, it can be assumed the yokes are not bent. Both must also be checked for cracks. If they are damaged or cracked, fit new replacements.

7 Fork legs: dismantling

1 Slacken and remove the top bolt, noting that it will be necessary to hold the stanchion whilst this is done. The easiest method is to clamp the leg in one of the fork yokes, or failing this, to use a strap wrench. It is not advisable to clamp the stanchion in a vice, because of the risk of damage due to scoring or distorting. Use of this method is permissible if soft jaws are used. Be very careful when unscrewing the top bolt. The fork spring is under tension and will push the bolt clear with some force. It follows that pressure should be applied to the bolt to counter this, firm hand pressure being adequate.

2 When the top bolt has been removed, invert the leg over a drain tray and leave it until the damping oil has drained. Repeat the above procedure on the remaining leg, but note that each leg should be dismantled and reassembled separately to avoid interchanging components.

3 Wrap some rag around the lower leg and clamp the assembly in a vice, taking care not to overtighten and thus distort the lower leg. Using an Allen key, slacken and remove the damper bolt from the recessed hole in the bottom of the lower leg. This will often cause some difficulty because the bolt threads are coated in a locking compound and once slightly loose there is a tendency for the damper rod to turn in its seat. To overcome this problem a length of wooden dowel can be employed to hold the damper rod. Grind a coarse taper on one end of the dowel and insert it down the bore of the stanchion so that it engages in the recessed head of the damper rod.

4 It will now be necessary to push on the dowel to obtain grip on the damper. The dowel must not be allowed to turn, and to this end it is recommended that a self-locking wrench is clamped across its end to provide a handle. With an assistant restraining the damper rod, the bolt should now unscrew. If working alone, cut the dowel so that it lies about $\frac{1}{2}$ in lower than the top of the fully extended stanchion. The top bolt can be temporarily refitted to apply pressure to the dowel whilst the bolt is removed.

5 Disengage and slide off the dust seal which is fitted around the top of the lower leg. The stanchion can now be withdrawn, taking care not to damage the seal lip. The damper seat can be tipped out of the lower leg and the damper and rebound spring removed from the stanchion in a similar manner.

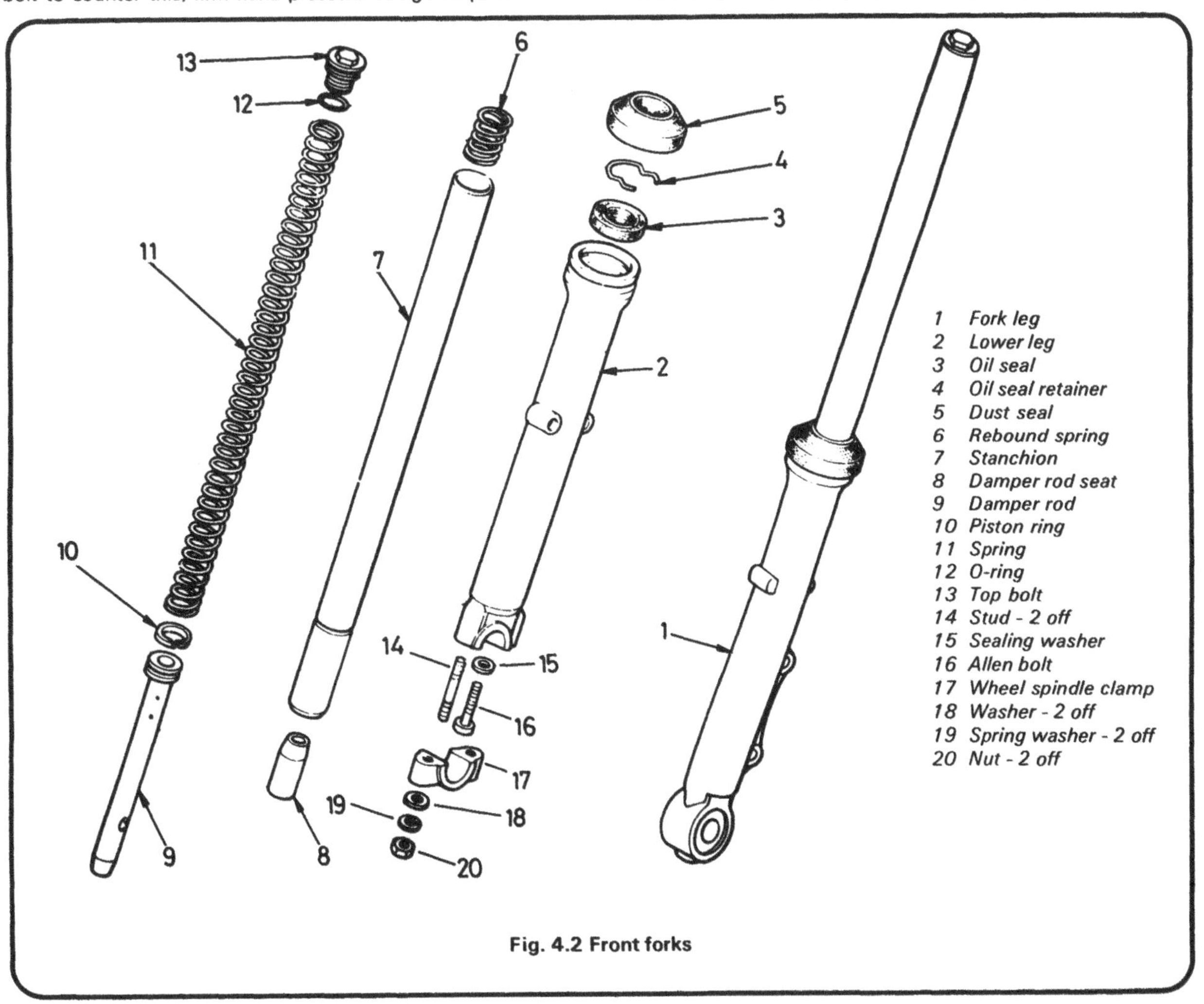

Fig. 4.2 Front forks

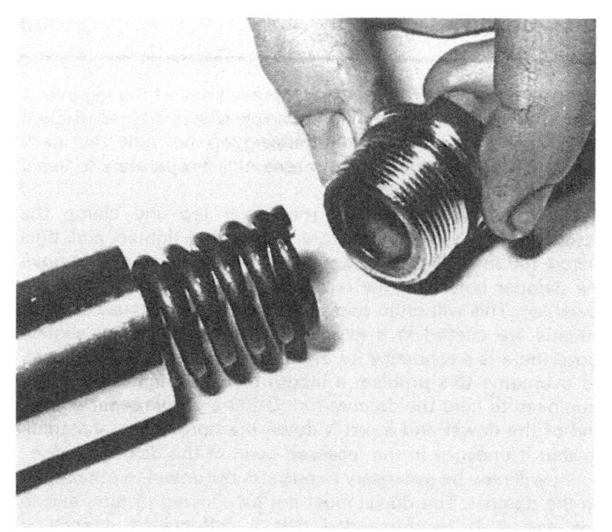
7.1a Remove top bolt noting that it is under spring pressure

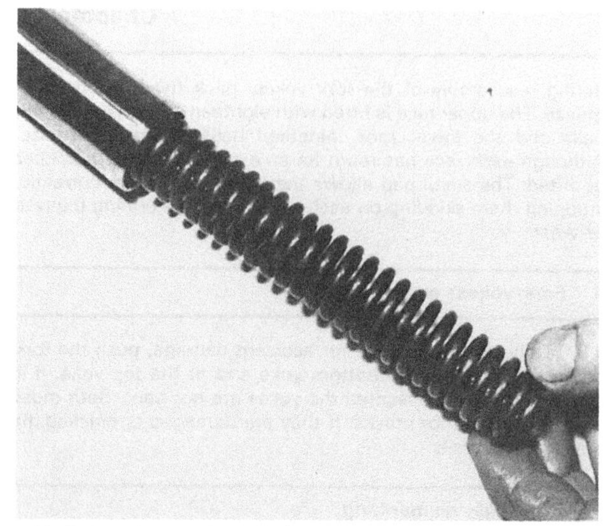
7.1b Fork spring can be withdrawn from stanchion

7.4 Remove damper lock bolt to free stanchion

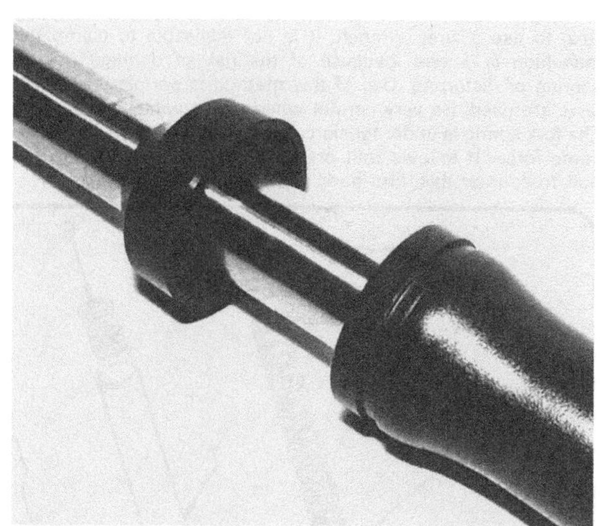
7.5a Slide off the dust seal ...

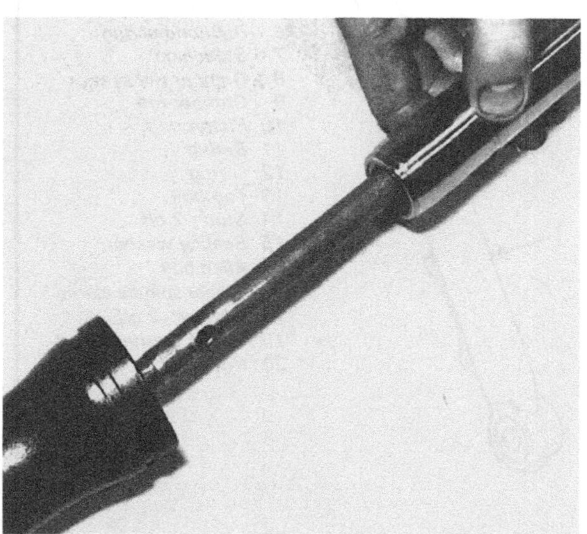
7.5b ... and withdraw stanchion and damper assembly

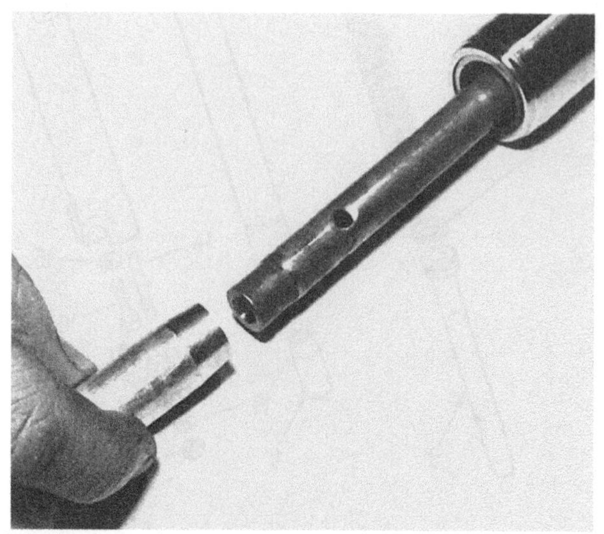
7.5c Damper rod seat is positioned as shown

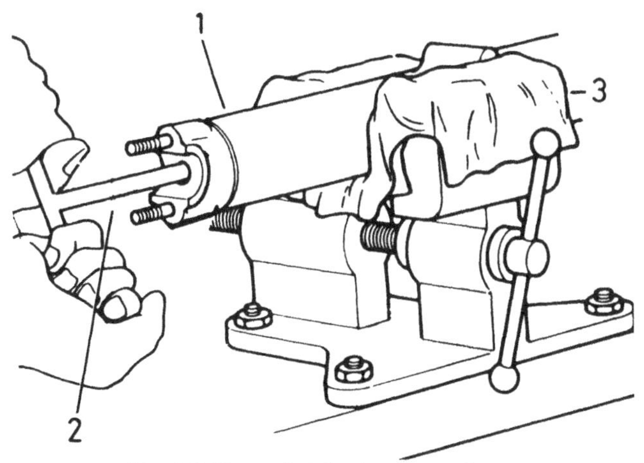

Fig. 4.3 Clamp fork leg in vice as shown

1 Fork lower leg
2 Allen bolt removing tool
3 Rag

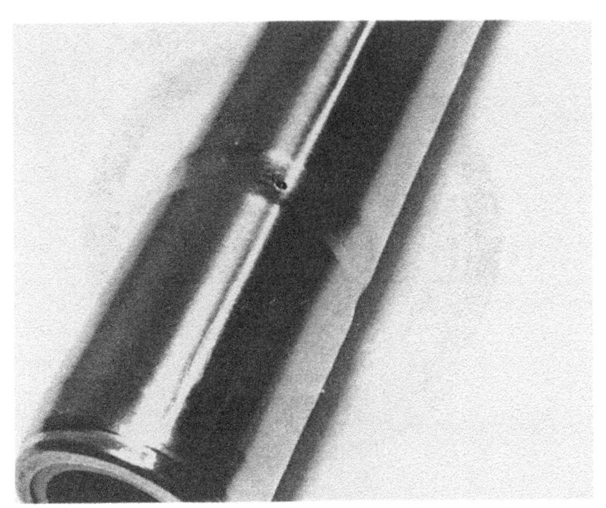

8.1a Check stanchion for scoring – note oil hole

8 Front forks: examination and renovation

1 The parts most likely to wear over an extended period of service are the internal surfaces of the lower leg and the outer surfaces of the fork stanchion or tube. If there is excessive play between these two parts they must be replaced as a complete unit. Check the fork tube for scoring over the length which enters the oil seal. Bad scoring here will damage the oil seal and lead to fluid leakage.

2 It is advisable to renew the oil seals when the forks are dismantled even if they appear to be in good condition. This will save a strip-down of the forks at a later date if oil leakage occurs. The oil seal in the top of each lower leg is retained by an internal C-ring which can be prised out of position with a small screwdriver. Check that the dust excluder rubbers are not split or worn where they bear on the fork tube. A worn excluder will allow the ingress of dust and water which will damage the oil seal and eventually cause wear of the fork tube.

3 It is not generally possible to straighten forks which have been badly damaged in an accident, particularly when the correct jigs are not available. It is always best to err on the side of safety and fit new ones, especially since there is no easy means to detect whether the forks have been over stressed or metal fatigued. Fork stanchions (tubes) can be checked, after removal from the lower legs, by rolling them on a dead flat surface. Any misalignment will be immediately obvious.

4 The fork springs will take a permanent set after considerable usage and will need renewal if the fork action becomes spongy. The length of the fork springs should be checked against the figures given in the Specifications Section.

5 The damping action of the forks is governed by the viscosity of the oil in the fork legs, and the recommended types and capacities are given in the Specifications at the beginning of this Chapter. Note that when the fork is being topped up or the damping oil changed, slightly less oil will be needed. It is recommended that a piece of wire is used as a dipstick in these cases, and the oil level measured from the top of the fork stanchion. Given that the initial oil capacity is correct it will be possible to translate this to a known oil level, and subsequent refills and experiments in oil level changes can be made on this basis.

6 It is possible to increase or reduce the damping effect by using a different oil grade, and some owners may wish to experiment a little to find a grade which suits their particular application. It is advisable to consult a Honda Service Agent who will be able to suggest which oils may be used.

8.1b Damper assembly can be tipped out of stanchion

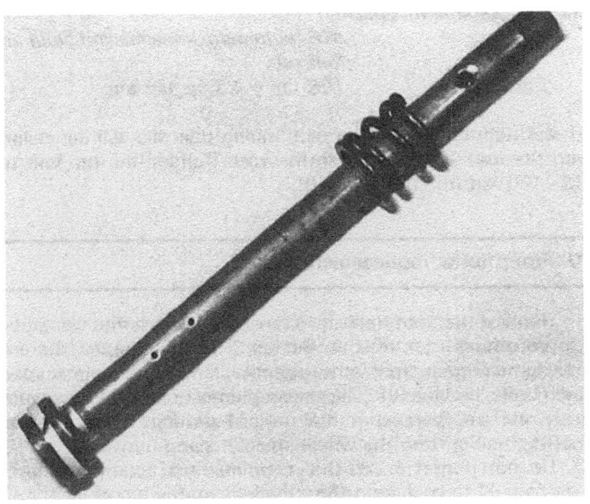

8.1c Check for corrosion and blocked oil drillings ...

8.2a Seal is retained by wire clip

8.2b Worn oil seals can be prised out as shown

9 Front fork legs: reassembly

1 All of the fork leg components should be completely clean and free from dust or oil prior to reassembly. Remember that the forks are in constant motion in use, and any abrasive particles will quickly wear away the surface against which they are trapped. If new seals are required they should be fitted at this stage. A large diameter socket can be used to drive the new seal into position. Lubricate the seal lip with grease.
2 Fit the damper assembly and rebound spring into the stanchion and drop the damper seat into the lower leg. Feed the stanchion into the lower leg, taking care not to damage the seal. Check that the damper bolt threads are clean and dry, then coat them with Loctite. Fit the damper bolt and tighten it to 1.5 - 2.5 kgf m (11 - 18 lbf ft).
3 Slide the dust seal into position, ensuring that it locates correctly. It is worthwhile wiping some grease around the outside of the oil seal before the dust seal is fitted. This will ensure that corrosion and scoring are prevented. Fit the fork spring, then top up each fork leg with the prescribed quantity and grade of damping oil.

Fork oil grade and capacity

Grade	*ATF (automatic transmission fluid) or fork oil*
Capacity	*158 cc ± 2.5 cc per leg*

Fit and tighten the fork top bolt, noting that this will be easier with the fork leg clamped in the yoke. Tighten the top bolt to 1.5 - 3.0 kgf m (11 - 22 lbf ft).

10 Front forks: replacement

1 Replace the front forks by following in reverse the dismantling procedures described in Section 3 of this Chapter. Before fully tightening the front wheel spindle clamps and the fork yoke pinch bolts, bounce the forks several times to ensure they work freely and are clamped in their original settings. Complete the final tightening from the wheel spindle clamp upwards.
2 Do not forget to add the recommended quantity of fork damping oil to each leg before the bolts in the top of each fork leg are replaced.
3 If the fork stanchions prove difficult to re-locate through the fork yokes, make sure their outer surfaces are clean and polished so that they will slide more easily. It is often advantageous to use a screwdriver blade to open up the clamps as the stanchions are pushed upward into position.
4 As the fork legs are fed through the yokes they will correct any slight misalignment between them should the steering head have been dismantled. Where appropriate, tighten the steering stem top nut and complete the fitting of the various ancillary components.
4 Before the machine is used on the road, check the adjustment of the steering head bearings. If they are too slack, judder will occur. There should be no detectable play in the head races when the handlebars are pulled and pushed, with the front brake applied hard.
5 Overtight head races are equally undesirable. It is possible to apply unwittingly a loading of several tons on the head bearings by overtightening, even though the handlebars appear to turn quite freely. Overtight bearings will cause the machine to roll at low speeds and give generally imprecise handling with a tendency to weave. Adjustment is correct if there is no perceptible play in the bearings and the handlebars will swing to full lock in either direction, when the machine is supported with the front wheel clear of the ground. Only a slight tap should cause the handlebars to swing.

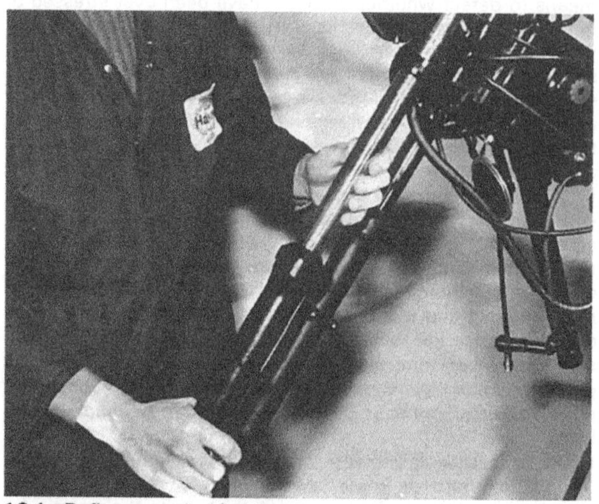
10.1a Refit assembled fork leg

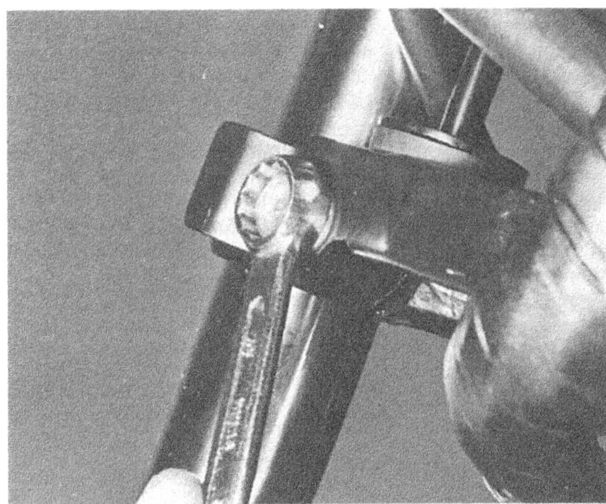
10.1b Tighten the lower yoke pinch bolt ...

10.1c ... and the upper yoke pinch bolt

10.2a Oil should be added whilst handlebar is removed ...

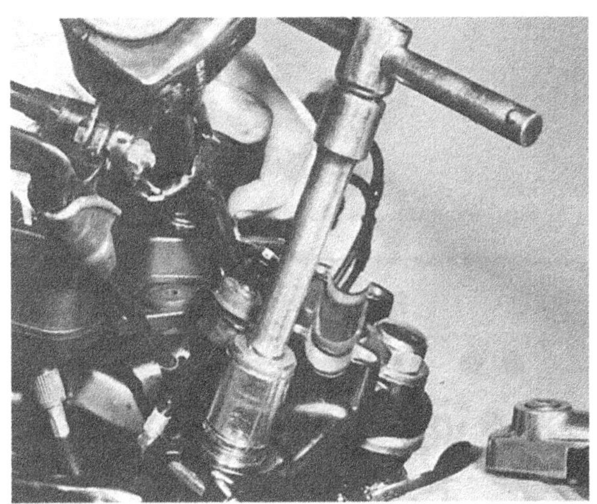
10.2b ... as should the fork top bolt

11 Frame: examination and renovation

1 The frame is unlikely to require attention unless accident damage has occurred. In some cases, replacement of the frame is the only satisfactory course of action if it is badly out of alignment. Only a few frame repair specialists have the jigs and mandrels necessary for resetting the frame to the required standard of accuracy and even then there is no easy means of assessing to what extent the frame may have been over-stressed.
2 After the machine has covered a considerable mileage, it is advisable to examine the frame closely for signs of cracking or splitting at the welded joints. Rust can also cause weakness at these joints. Minor damage can be repaired by welding or brazing, depending on the extent and nature of the damage.
3 Remember that a frame which is out of alignment will cause handling problems and may even promote speed wobbles. If misalignment is suspected, as the result of an accident, it will be necessary to strip the machine completely so that the frame can be checked and, if necessary, renewed.

12 Rear suspension unit: removal and examination

1 The models featured in this manual are equipped with coil spring suspension units using oil-filled damper assemblies. The units are mounted at a steep angle to provide maximum rear wheel travel.
2 It is best to remove and attend to one unit at a time, because this will allow the machine to be supported by the remaining unit. Alternatively, place the machine on its centre stand so that the rear wheel is raised clear of the ground and the weight taken from the suspension units.
3 The units are retained by a single mounting bolt at each end. With the bolts removed, the units can be pulled clear of the frame.
4 Each unit can be dismantled after the spring has been compressed sufficiently for the lower seat to be disengaged and displaced, a slot providing clearance to enable it to be slid around the damper rod. Ideally, this should be done by compressing the unit with the appropriate spring compressor. It was found in practice that one person could compress the

spring adequately by hand, whilst an assistant removed the lower spring seat. If the latter method is chosen, great care must be taken to avoid injury or damage as a result of the assembly slipping under tension.

5 Measure the free length of the spring, and renew it if it is below the service limit specified. The damper unit is sealed and cannot be dismantled. Its operation can be checked to some extent by compressing it and then releasing it. The unit should show significant damping effect on rebound (extension) but much less under compression. Any signs of leakage will necessitate renewal of the damper units as a pair.

6 If the suspension units appear to be in need of renewal, thought should be given to fitting an improved proprietary replacement pair. These may prove to be more expensive, but usually provide better control and durability. Most accessory shops will be able to advise and recommend the best make and type to use for any given application.

7 Reassembly of the units is a straightforward reversal of the dismantling sequence. When fitting the spring adjuster, set it at its safest position to ease spring compression. When the units have been refitted set the spring adjusters at the required setting, ensuring that the position of each adjuster matches the other.

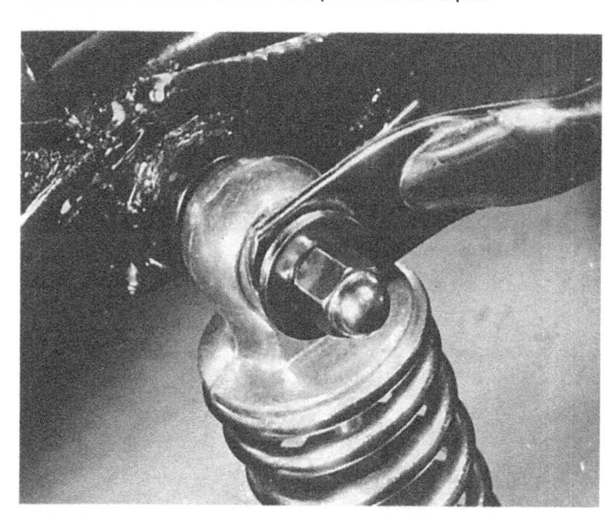

12.3a Suspension units are secured by domed nut at top ...

12.3b ... and by bolt at bottom

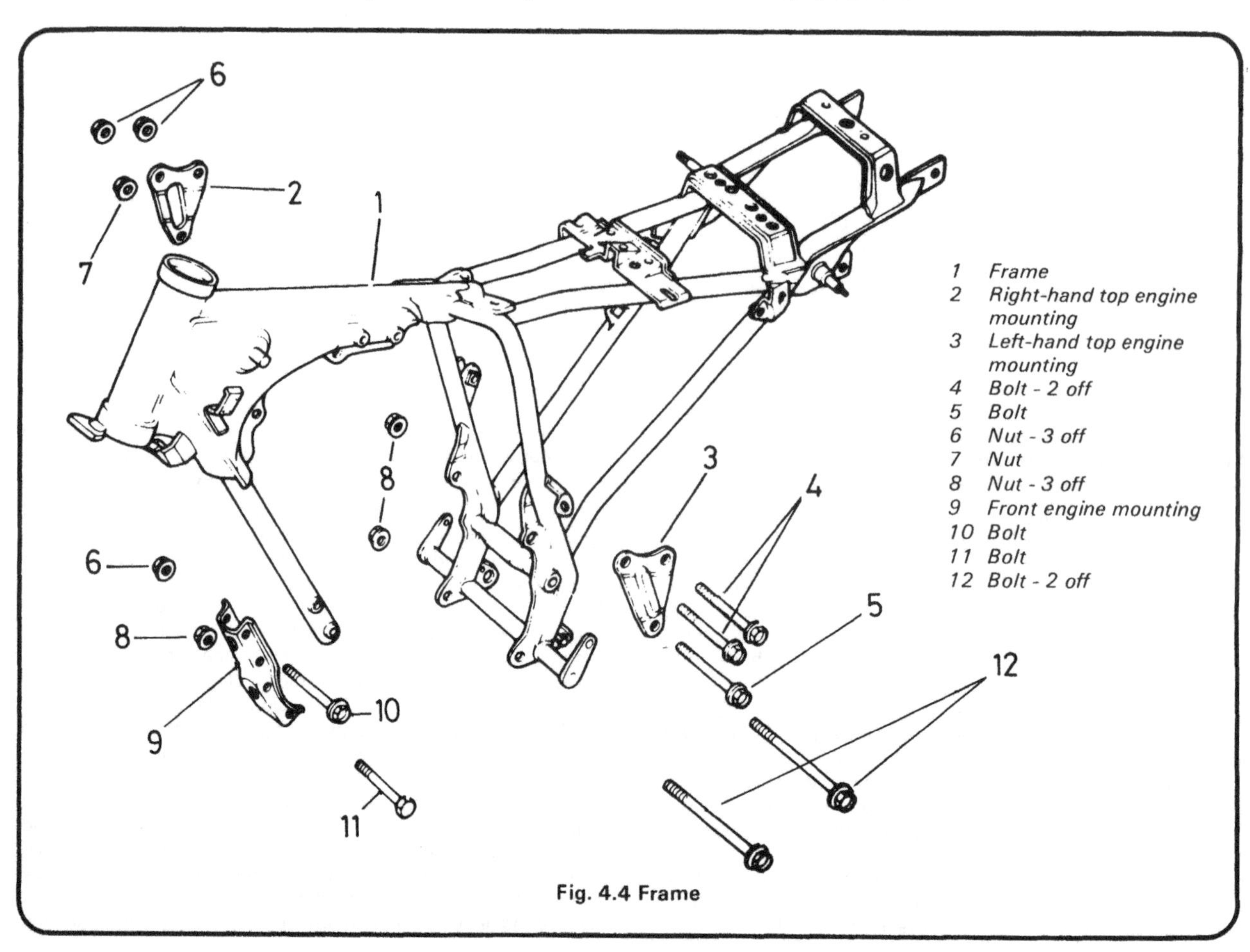

Fig. 4.4 Frame

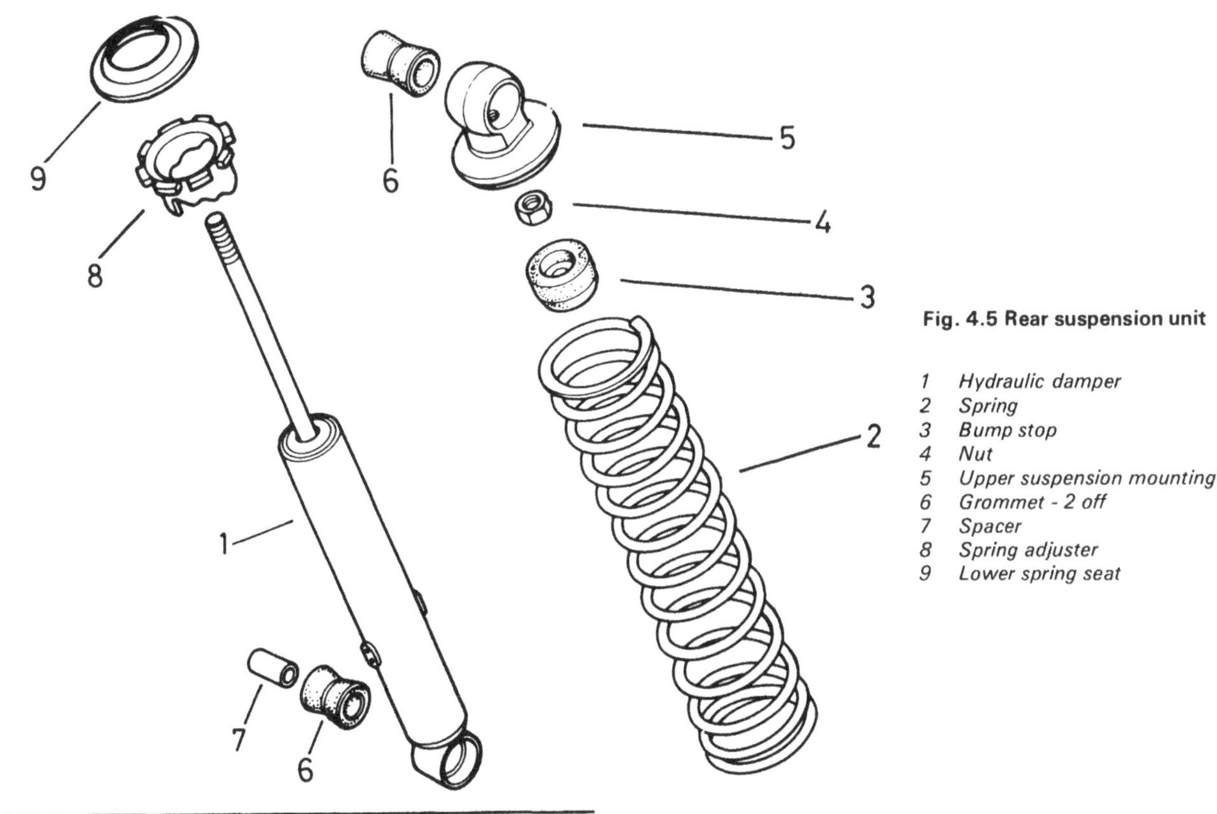

Fig. 4.5 Rear suspension unit

1 Hydraulic damper
2 Spring
3 Bump stop
4 Nut
5 Upper suspension mounting
6 Grommet - 2 off
7 Spacer
8 Spring adjuster
9 Lower spring seat

13 Swinging arm rear forks: dismantling, examination and renovation

1 The swinging arm fork pivots on a pair of headed bushes which are pressed into the fork crossmember. The internal surfaces of the bushes bear upon a hard steel sleeve, which in turn is secured by a long pivot bolt between the frame bosses.

2 Wear will inevitably take place in the bushes after a period of time, although regular greasing, using the grease nipple provided, will greatly extend their life. If greasing gets put off indefinitely, as often happens, the bearing surfaces will dry up allowing rapid wear to take place.

3 One of the first symptoms of wear is a sensation of vagueness in the machine's handling. Curiously, it is often the front forks which are blamed initially, as the problem seems to be one of imprecise steering. Pronounced wear will allow the swinging arm and rear wheel to jump from one side of the extent of play to the other. Such a machine is very far from roadworthy, and should not be used until an overhaul has taken place.

4 An indication of wear can be found by placing the machine on its centre stand so that the rear wheel is clear of the ground. If the fork is pushed and pulled horizontally, any play in the bushes will be proportionally exaggerated at the rear of the fork. Any discernible play will warrant immediate renewal of the worn parts.

5 Remove the rear wheel as described in Chapter 5, Section 12, then release the plastic chainguard by removing its two retaining screws.

6 Remove the lower mounting bolt from each of the rear suspension units, which can then be pivoted upwards to clear the swinging arm. Slacken and remove the pivot shaft nut. Support the swinging arm and withdraw the spindle. As the swinging arm is drawn away from the frame the end caps will probably drop away, and they should be retrieved and placed to one side.

7 Displace the hardened steel sleeve, and wash the swinging arm bore, bushes and sleeve in clean petrol (gasoline) or degreasing solution. Check the internal bore of the bushes for wear or scoring, and likewise the steel sleeve for wear and corrosion. The bushes should be renewed as a matter of course if any trace of movement has been detected. Note that the bushes will certainly be damaged during removal, and therefore cannot be reused once they have been driven out of the swinging arm.

8 The bushes can be removed by passing a long drift or bar through the swinging arm bore, and driving out the bush on the opposite side. Take care to support the swinging arm to avoid any risk of damage. When fitting new bushes, place a flat plate across the head of the bush to spread the impact evenly, and ensure that the bush enters the bore squarely.

9 Reassemble the swinging arm by reversing the dismantling sequence. The steel sleeve should be coated with high melting point grease. Do not omit to refit the end caps to the swinging arm before installing it in the frame. The end caps can be held in position with grease, or by passing a screwdriver into the bore through the frame lug. Tighten the swinging arm pivot shaft nut to 6.0 - 8.0 kgf m (43 - 58 lbf ft).

14 Stands: examination and maintenance

1 The prop stand, or side stand, consists of a steel leg attached to the left-hand side of the frame by means of a pivot bolt which passes through a frame lug. An extension spring holds the stand in a horizontal, retracted position when not in use.

2 Check that the pivot bolt is secured and that the extension spring is in good condition and not overstretched. An accident is almost certain if the stand extends whilst the machine is on the move.

3 The centre stand is secured by a tubular, headed pivot shaft which is retained in turn by a split pin. The stand should be removed occasionally, and the pivot shaft cleaned and re-greased. Check that the return spring is in good condition and not likely to fail whilst the machine is in use. If stretched, cracked or badly rusted, the spring should be renewed.

Fig. 4.6 Swinging arm

1 *Swinging arm*
2 *Grease nipple - 2 off*
3 *Headed bush - 4 off*
4 *Steel sleeve - 2 off*
5 *End cap - 4 off*

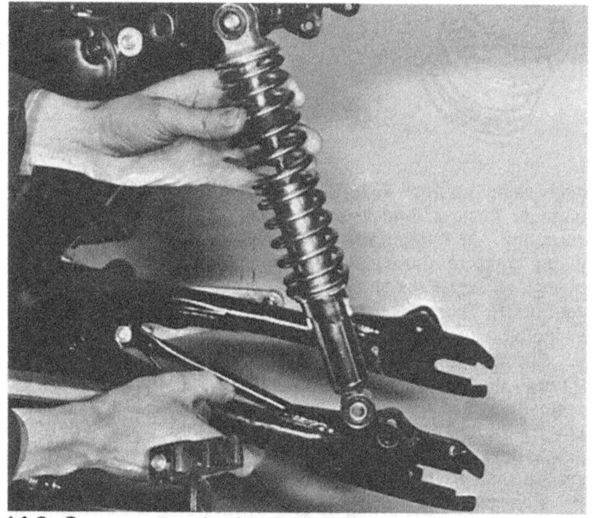

14.6a Remove rear wheel and suspension units

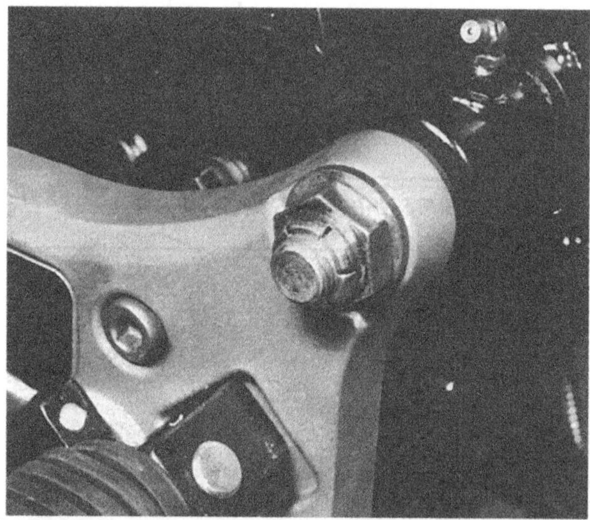

14.6b Slacken self-locking nut at swinging arm pivot

14.6c Slacken mounting plate(s) if necessary

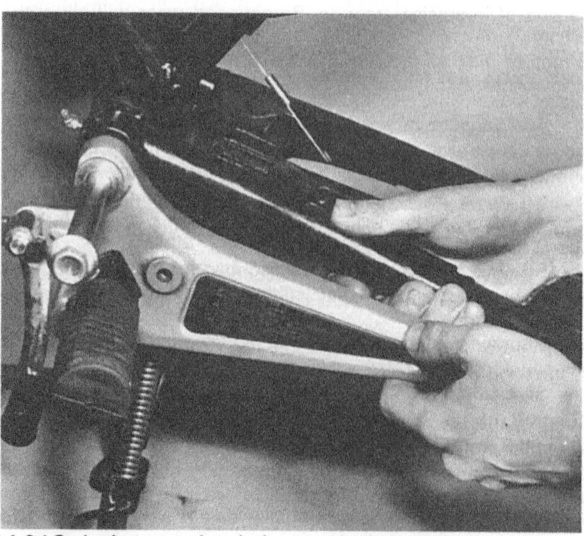

14.6d Swinging arm pivot bolt can now be removed

14.7a Displace end caps from fork ends ...

14.7b ... and push out hardened steel sleeves

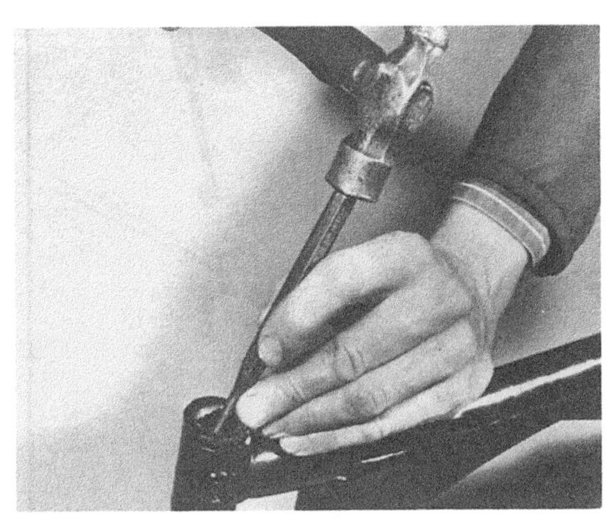

14.8 Bushes can be driven out as shown

15 Footrests: examination and renovation

1 The footrests are attached to lugs which are in turn bolted to the footrest plate mounted on each side of the machine. All four footrests are of the folding type which will minimise damage to them should the machine be dropped, though offering less protection to the remaining cycle parts than the rigid types.

2 If the footrests are damaged in an accident, it is possible to dismantle the assembly into its component parts. Detach each footrest from the frame lugs and separate the folding foot piece from the bracket on which it pivots by withdrawing the split pin and pulling out the pivot shaft. It is preferable to renew the damaged parts, but if necessary, they can be bent straight by clamping them in a vice and heating to a dull red with a blow lamp whilst the appropriate pressure is applied. Do not attempt to straighten the footrests while they are attached to the frame.

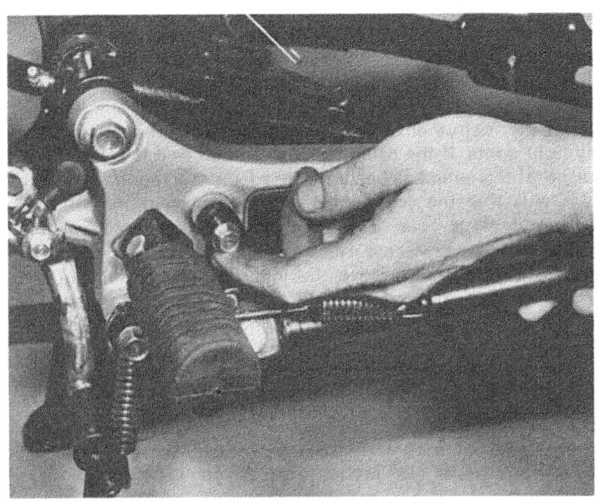

15.1 Side stand assembly bolts to footrest plate

16 Rear brake pedal: removal and examination

1 The rear brake pedal is clamped to the splined end of a pivot shaft which runs in the right-hand footrest plate. The pedal may be removed by releasing the clamp bolt and sliding it off its splines. If the pedal becomes bent it can be straightened using the technique described in Section 15.

2 The pivot shaft should be displaced and re-greased whenever the pedal is removed. This will prevent wear in the bore through which it passes and will stop dried up grease from causing problems.

17 Speedometer and tachometer heads: removal

1 The speedometer and tachometer heads are incorporated in a moulded plastic instrument panel which is attached to the upper fork yoke. If either instrument malfunctions, it will be necessary to remove the instrument panel and separate the two halves to gain access to the individual instrument heads.

2 Disconnect the speedometer and tachometer drive cables by unscrewing the knurled ring which retains each one. Slacken the two rubber-mounted bolts that pass through the panel on either side of the ignition switch. Lift the assembly up and pull out each of the lamp holders from the underside. Release the ignition switch by removing the two retaining bolts. The panel can now be lifted clear and taken to the workbench for dismantling.

3 Remove the recessed cross-head screws from the underside of the panel to free the upper half, having released the trip

reset knob by unscrewing its single retaining screw. The instrument heads will now be exposed and can be freed by releasing the retaining screws which secure them to the lower panel half.

4 The instrument heads are very delicate and should not be dismantled at home. In the event of a fault developing, the instrument should be entrusted to a specialist repairer or a new unit fitted. If a replacement unit is required it is well worth trying to obtain a good secondhand item from a motorcycle breaker in view of the high cost of a new instrument.

5 Remember that a speedometer in correct working order is a statutory requirement in the UK. Apart from this legal necessity, reference to the odometer readings is the most satisfactory means of keeping pace with the maintenance schedules.

18 Speedometer and tachometer drive cables: examination and maintenance

1 If the operation of the speedometer or tachometer becomes sluggish or jerky it can often be attributed to a damaged or kinked drive cable. Remove the cable from the machine by releasing the locating screw at the lower end and the knurled ring at the instrument head. Turn the inner cable to check for any tight spots. If the cable tends to snatch as it is rotated it is likely that the inner has become kinked and will require renewal.

2 Check that the outer cable is in sound condition with no obvious damage. The inner and outer cables are available separately where necessary. When fitting a new cable or refitting an old one, remove the inner and grease all but the upper six inches or so. This will ensure that the cable is adequately lubricated without incurring the risk of grease working up into the instrument head. Ensure that the cable is routed in a smooth path between the drive gearbox and instrument panel, avoiding any tight bends which would result in the cable becoming kinked.

19 Speedometer and tachometer drives: location and examination

1 The speedometer cable is driven by a worm mechanism incorporated in the front brake plate. The assembly is of robust construction and requires no regular maintenance. Apart from obvious breakages, which will require renewal of the parts concerned it will suffice to grease the drive mechanism whenever attention is given to the front wheel bearings.

2 The tachometer is driven by a gearbox mounted on the end of the camshaft. This too is unlikely to require regular attention, but can be greased whenever the cylinder head receives attention.

20 Cleaning the machine

1 After removing all surface dirt with a rag or sponge which is washed frequently in clean water, the machine should be allowed to dry thoroughly. Application of car polish or wax to the cycle parts will give a good finish, particularly if the machine receives this attention at regular intervals.

2 The plated parts should require only a wipe with a damp rag, but if they are badly corroded, as may occur during the winter when the roads are salted, it is permissible to use one of the proprietary chrome cleaners. These often have an oily base which will help to prevent corrosion from recurring.

3 If the engine parts are particularly oily, use a cleaning compound such as Gunk or Jizer. Apply the compound whilst the parts are dry and work it in with a brush so that it has an opportunity to penetrate and soak into the film of oil and grease. Finish off by washing down liberally, taking care that water does not enter the carburettor, air cleaner or any of the electrical connectors.

4 If possible, the machine should be wiped down immediately after it has been used in the wet, so that it is not garaged under damp conditions which will promote rusting. Make sure that the chain is wiped and re-oiled, to prevent water from entering the rollers and causing harshness with an accompanying rapid rate of wear. Remember there is less chance of water entering the control cables and causing stiffness if they are lubricated regularly as described in the Routine Maintenance Section.

5 The matt black finish on the engine casings will tend to wear through or chip over a period of time. It can be restored with one of the proprietary aerosol paints, but care must be taken to ensure that a heat resistant paint is chosen.

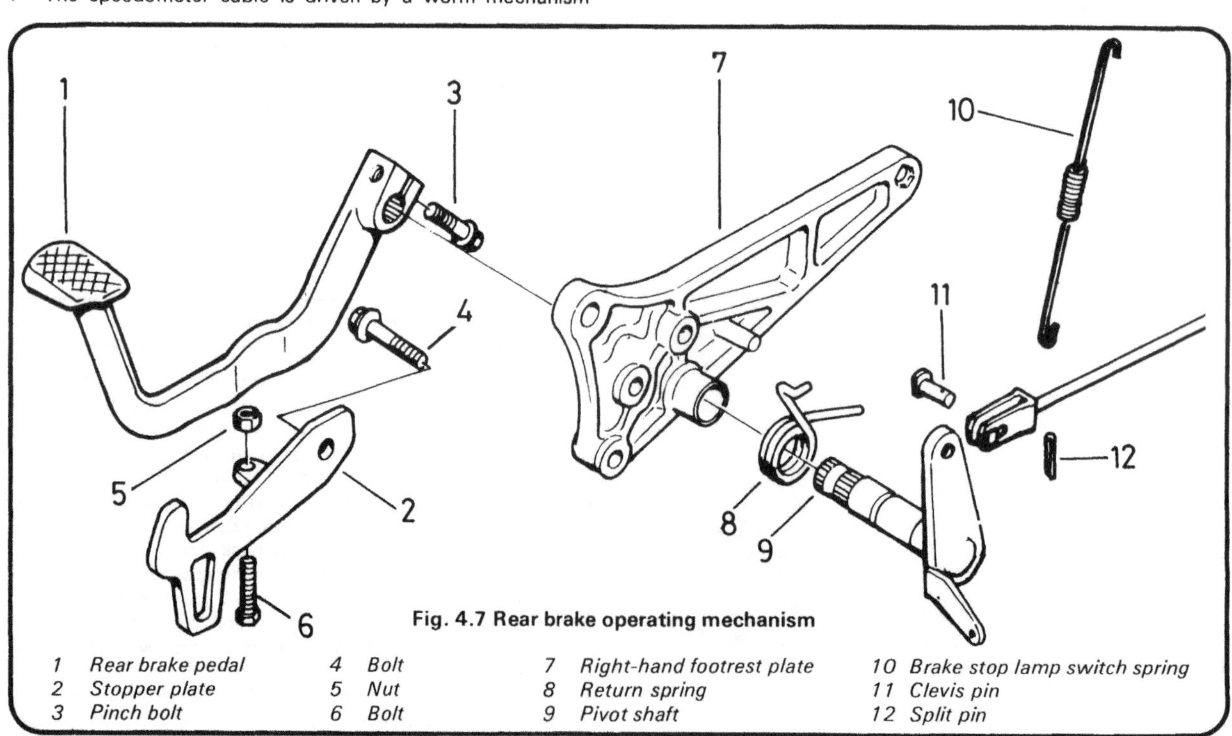

Fig. 4.7 Rear brake operating mechanism

1	*Rear brake pedal*	*4*	*Bolt*	*7*	*Right-hand footrest plate*	*10*	*Brake stop lamp switch spring*
2	*Stopper plate*	*5*	*Nut*	*8*	*Return spring*	*11*	*Clevis pin*
3	*Pinch bolt*	*6*	*Bolt*	*9*	*Pivot shaft*	*12*	*Split pin*

18.2a Remove bolts to free instrument panel

18.2b Remove ignition switch and wiring

18.3a Release trip reset knob (single screw)

18.3b Remove casing screws...

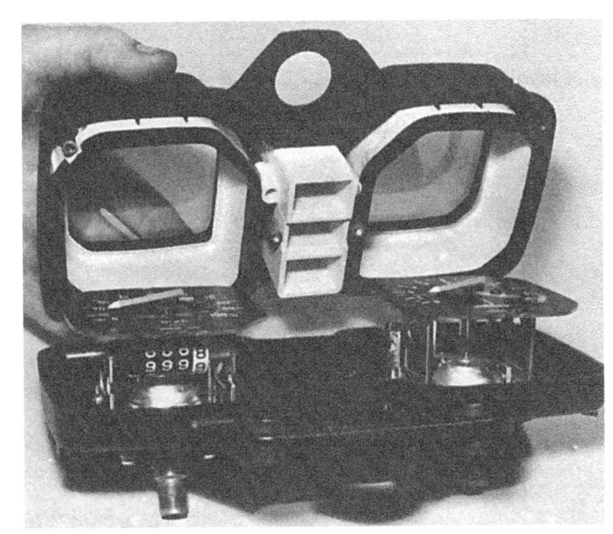

18.3c ... and lift upper casing away

18.3d Instruments are retained by two screws

18.4 DO NOT attempt to dismantle mechanism at home

20.2 Tachometer drive is a self-contained unit

21 Fault diagnosis: frame and forks

Symptom	Cause	Remedy
Machine veers either to the left or the right with hands off handlebars	Bent frame Twisted forks Wheels out of alignment	Check and renew. Check and renew. Check and re-align.
Machine rolls at low speed	Overtight steering head bearings	Slacken until adjustment is correct.
Machine judders when front brake is applied	Slack steering head bearings Worn fork legs	Tighten until adjustment is correct. Dismantle forks and renew worn parts.
Machine pitches on uneven surfaces	Ineffective fork dampers Ineffective rear suspension units Suspension too soft	Check oil content. Check whether units still have damping action. Raise suspension unit adjustment one notch.
Fork action stiff	Fork legs out of alignment (twisted in yokes)	Slacken yoke clamps, and fork top bolts. Pump fork several times then retighten from bottom upwards.
Machine wanders. Steering imprecise. Rear wheel tends to hop	Worn swinging arm pivot	Dismantle and renew bushes and pivot shaft.

Note: *Tyre pressures and wheel condition can also affect handling and roadholding. Refer to Chapter 5 for details*

Chapter 5 Wheels, brakes and tyres

For information relating to the CB250 RSD-C model, refer to Chapter 7

Contents

Specifications

Wheels		
Type	Wire spoked, aluminium rims	
Brakes		
Front	Hydraulically operated single disc	
Disc thickness	4.9 – 5.1 mm (0.19 – 0.20 in)	
Service limit	4.0 mm (0.16 in)	
Rear	Single leading shoe	
Drum diameter	140 mm (5.51 in)	
Service limit	141 mm (5.55 in)	
Lining thickness	4 mm (0.16 in)	
Service limit	2 mm (0.08 in)	
Tyres		
Front	3.00S18-4PR	
Rear	4.10S18-4PR	
Tyre pressures	**Front**	**Rear**
Solo	24 psi (1.75 kg/cm^2)	32 psi (2.25 kg/cm^2)
With pillion	24 psi (1.75 kg/cm^2)	36 psi (2.50 kg/cm^2)
Torque wrench settings	**kgf m**	**lbf ft**
Front wheel spindle nut	5.0 – 8.0	36 – 58
Front spindle clamp nuts	1.8 – 2.5	13 – 18
Front brake disc bolts	2.7 – 3.8	20 – 27
Brake caliper bolts	3.0 – 4.5	22 – 33
Rear wheel spindle nut	8.0 – 10.0	60 – 72
Rear brake torque arm bolts	1.8 – 2.5	13 – 18
Sprocket bolt nuts	6.0 – 7.0	43 – 51

1 General description

The Honda CB250 RS employs wheels of wire spoked construction, using light alloy 18 inch wheel rims front and rear. The front wheel is fitted with a 3.00S18-4PR tyre, whilst the rear rim carries a 4.10S18-4PR tyre.

A single hydraulically-operated front disc brake is fitted, rear wheel braking being by means of a rod-operated single leading shoe (sls) drum brake.

2 Front wheel: examination and renovation

1 Place the machine on the centre stand so that the front wheel is raised clear of the ground. Spin the wheel and check the rim alignment. Small irregularities can be corrected by tightening the spokes in the affected area although a certain amount of experience is necessary to prevent over-correction. Any flats in the wheel rim will be evident at the same time. These are more difficult to remove and in most cases it will be

necessary to have the wheel rebuilt on a new rim. Apart from the effect on stability, a flat will expose the tyre bead and walls to greater risk of damage if the machine is run with a deformed wheel.
2 Check for loose and broken spokes. Tapping the spokes is the best guide to tension. A loose spoke will produce a quite different sound and should be tightened by turning the nipple in an anti-clockwise direction. Always check for run out by spinning the wheel again. If the spokes have to be tightened by an excessive amount, it is advisable to remove the tyre and tube as detailed in Section 19 of this Chapter. This will enable the protruding ends of the spokes to be ground off, thus preventing them from chafing the inner tube and causing punctures.

3 Front wheel: removal and refitting

1 Place the machine on its centre stand, arranging blocks beneath the crankcase to raise the front wheel clear of the ground. Remove the single screw which retains the speedometer drive cable at the drive gearbox, lodging the cable clear of the wheel.
2 Straighten the split pin which retains the wheel spindle nut, withdrawing it from the spindle with a pair of pliers. Slacken and remove the castellated nut. Moving to the right-hand side of the wheel, remove the wheel spindle clamp nuts and displace the clamp. Support the wheel and withdraw the spindle. The wheel can now be lowered and manoeuvred clear of the forks. It is recommended that a wooden wedge is fitted between the brake caliper pads. If this precaution is ignored and the front brake lever is inadvertently operated, the pad and piston will be expelled and the caliper will require overhauling and bleeding.
3 The wheel is fitted in the reverse of the removal order, noting the following points. Ensure that the clamp half is fitted with the arrow mark facing forward. Fit the clamp nuts finger tight, then tighten the **front** nut fully. The rear clamp nut can now be secured. Note the torque settings shown below for the clamp and wheel spindle nuts.

Torque settings

Clamp nuts	*1.8 – 2.5 kgf-m (13.0 – 18.0 lbf ft)*
Spindle nut	*5.0 – 8.0 kgf-m (36.0 – 58.0 lbf ft)*

4 Front disc brake: general

1 As already mentioned, the front brake is of the hydraulically-operated disc type. It should be noted that a number of precautions should be taken when dealing with this system, because brake failure can have disastrous consequences.
2 The hydraulic system must be kept free from air bubbles. Any air in the system will be compressed when the brake lever is operated instead of transmitting braking effort to the disc. It follows that efficiency will be impaired, and given sufficient air, can render the brake inoperative. If any part of the hydraulic system is disturbed, the system must always be bled to remove any air. See Section 9 for details. It is vital that all hoses, pipes and unions are examined regularly and renewed if damage, deterioration or leakage is suspected.
3 Hydraulic fluid is specially formulated for given applications and must always be of the correct type. Any fluid conforming to SAE J1703 or DOT 3 may be used; other types may not be suitable. On no account should any other type of oil or fluid be used. Old or contaminated fluid must be discarded. It is dangerous to use old fluid which may have degraded to the point where it will boil in the caliper creating air bubbles in the system. Note that brake fluid will attack and discolour paintwork and plastics. Care must be taken to avoid contact and any accidental splashes washed off immediately.
4 Cleanliness is more important with hydraulic systems than in any other single area of the motorcycle. Dirt will rapidly destroy seals, allowing fluid to leak out or air to be drawn in. Water, even in the form of moist air, will be absorbed by the fluid which is hygroscopic. Fluid degraded by water has a lowered boiling point and can boil in use. The master cylinder and any cans of fluid must be kept securely closed to prevent this.

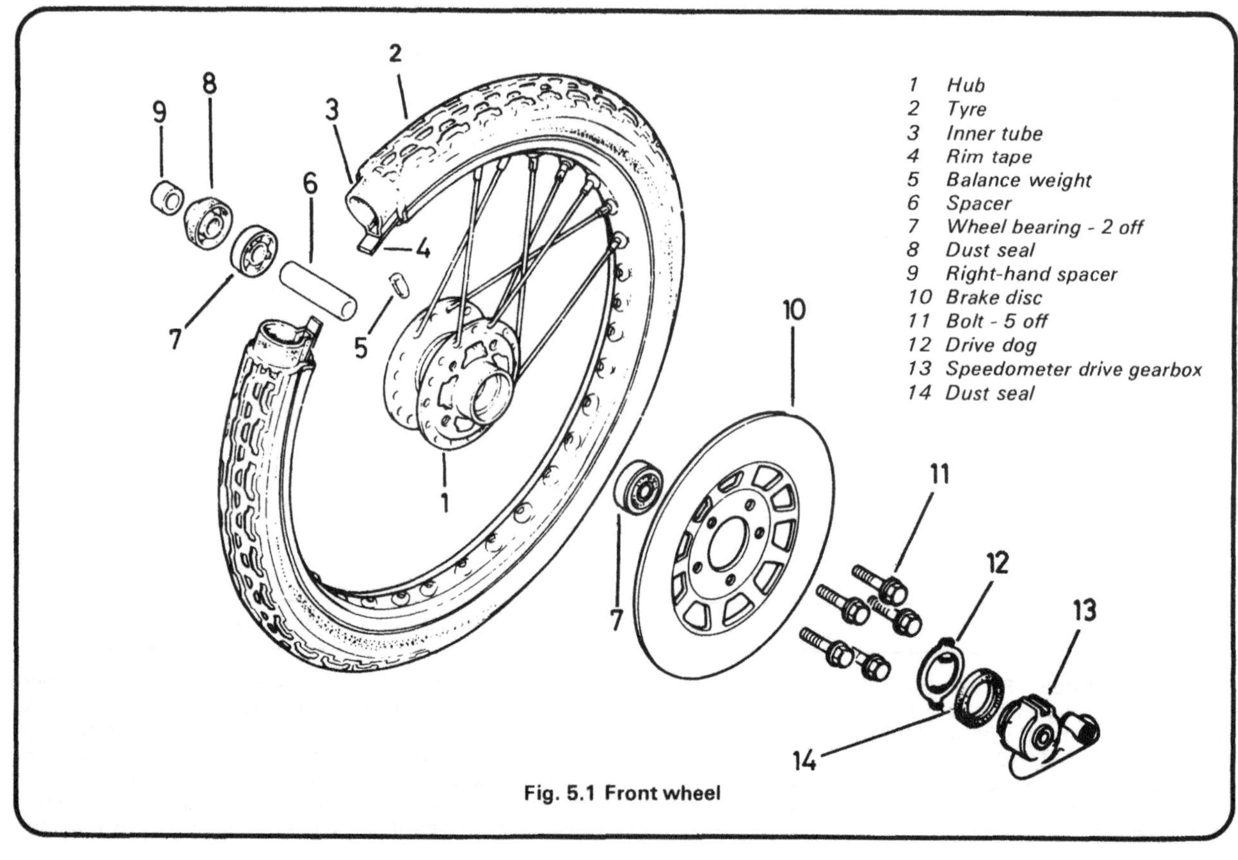

Fig. 5.1 Front wheel

5 Front disc brake: examination and renovation

1 Check the front brake master cylinder, hose and caliper unit for signs of fluid leakage. Pay particular attention to the condition of the hose, which should be renewed without question if there are signs of cracking, splitting or other exterior damage.
2 Check the level of hydraulic fluid by viewing through the translucent reservoir. If the fluid is below the lower level mark, with the handlebars so placed that the reservoir is vertical, brake fluid of the correct grade should be added. The correct fluid should conform to SAE J1703 (UK) or DOT 3 (USA) specifications. **Never use engine oil** or any fluid other than that recommended. Other fluids have unsatisfactory characteristics and will rapidly destroy the seals.
3 The brake pads should be inspected for wear. Each has a red groove which marks the limit of the friction material. When this limit is reached, **both** pads must be renewed, even if only one has reached the wear mark.
4 The pad wear marks may be viewed from the rear of the caliper assembly after removing the plastic inspection cover. If the pads require renewal they may be removed from the caliper without removing the wheel as follows: remove the single screw which holds the fluted cover to the caliper. Lift off the cover to expose the pads. Prise out the pad locating pin spring clip. Before withdrawing the pins, push the caliper hard over to the right. This will cause the piston to retreat into the cylinder, allowing room for the subsequent fitting of new pads.
5 After removing the two pins, the pads may be lifted from position individually. Note the shim placed against the rear face of the outer pad. Refit the new pads and the shim by reversing the dismantling procedure. Ensure that after fitting the pad pins, the retaining spring is installed so that each arm enters the drilled radial hole in the appropriate pin.

6 Front disc brake: overhauling the caliper unit

1 The caliper unit may be removed from the fork leg either with the front wheel in position or after the wheel has been removed. Remove the brake pads as described in the previous Section.
2 Before continuing dismantling the caliper, the brake fluid should be drained. Disconnect the brake pipe at the connection with the caliper by removing the banjo bolt, and allow the fluid to drain into a suitable container. Pump the handlebar lever gently until all the fluid has been expelled.
3 Remember that brake fluid is an excellent paint stripper; do not allow it to come into contact with any painted surface, or with any clear plastic such as is sometimes used for instrument glasses.
4 Unscrew the two bolts which pass through the caliper unit, and upon which the unit slides during operation. The two halves of the unit can now be separated. Push out the bolts to free the inner caliper half from the support bracket. The support bracket is secured to the fork leg by two bolts.
5 Remove the piston boot from the caliper cylinder (outer caliper half). To displace the piston, apply a blast of compressed air to the fluid inlet. Take care to catch the piston as it emerges – if dropped or prised out with a screwdriver the piston may be irreparably damaged.
6 The parts removed should be cleaned thoroughly, using only brake fluid as the liquid. Petrol, oil or paraffin will cause the various seals to swell and degrade, and should not be used under any circumstances. When the various parts have been cleaned, they should be stored in polythene bags until re-assembly, so that they are kept dust free.
7 Examine the piston for score marks or other imperfections. If it has any imperfections it must be renewed, otherwise air or hydraulic fluid leakage will occur, which will impair braking efficiency. With regard to the various seals, it is advisable to renew them all, irrespective of their appearance. It is a small price to pay against the risk of a sudden and complete front brake failure. Check the slider bolts for wear, together with the holes in the support bracket in which they slide. Wear at these points will cause brake judder and poor brake release.
8 Reassemble under clinically-clean conditions, by reversing the dismantling procedure. Apply a small quantity of graphite grease to the slider bolts before fitting the boots. Reconnect the hydraulic fluid pipe and make sure the union has been tightened fully. Before the brake can be used, the whole system must be refilled with brake fluid and bled of air, by following the procedure described in Section 9 of this Chapter.

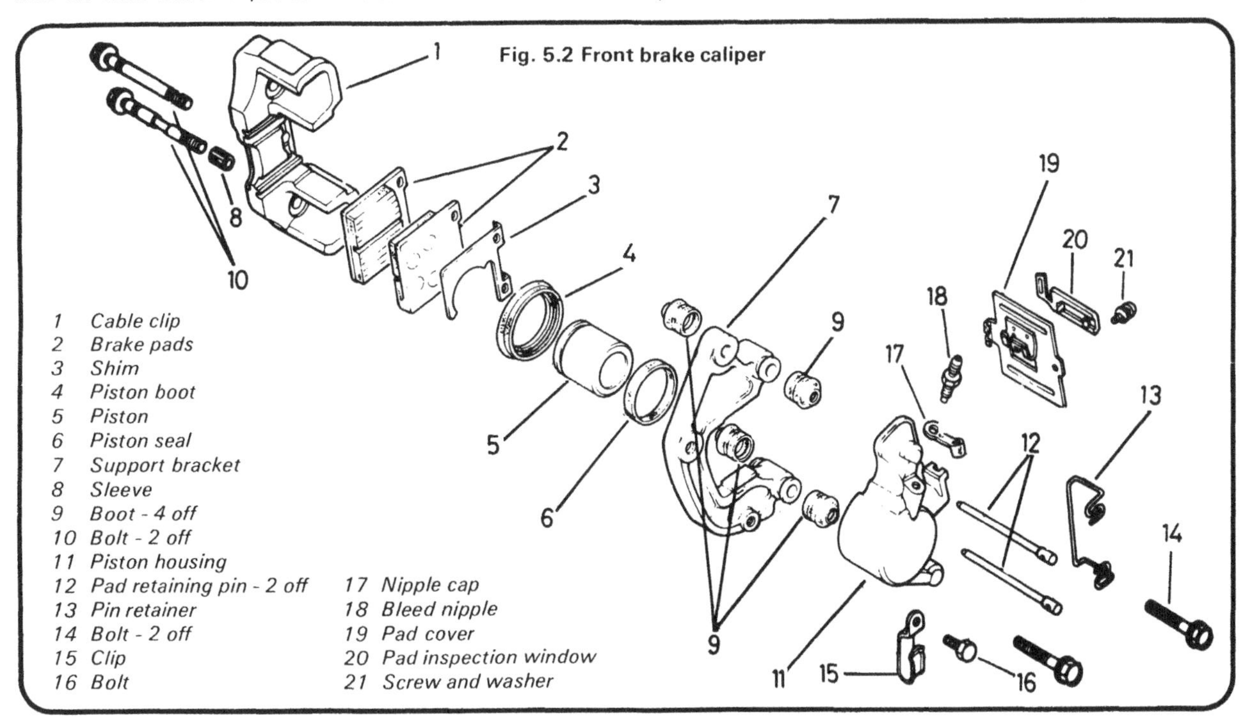

Fig. 5.2 Front brake caliper

3.1 Release the speedometer drive cable

3.2 Wheel spindle end is secured by half clamp

5.4a Remove screw (arrowed) to free brake pad cover

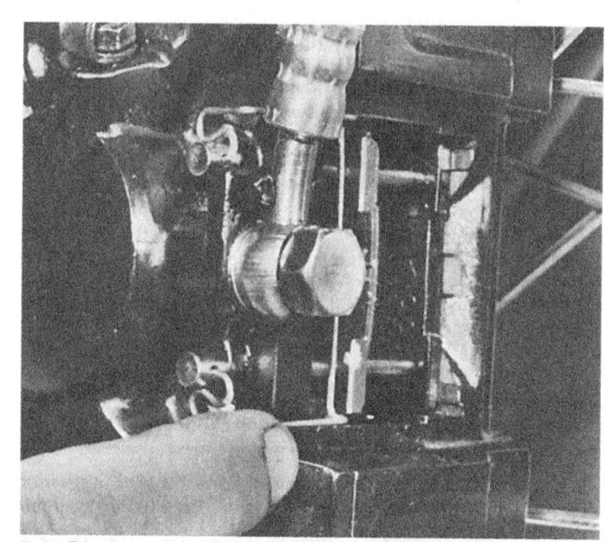
5.4b Displace spring wire clip from pin ends

5.4c Withdraw pins from caliper and pad holes

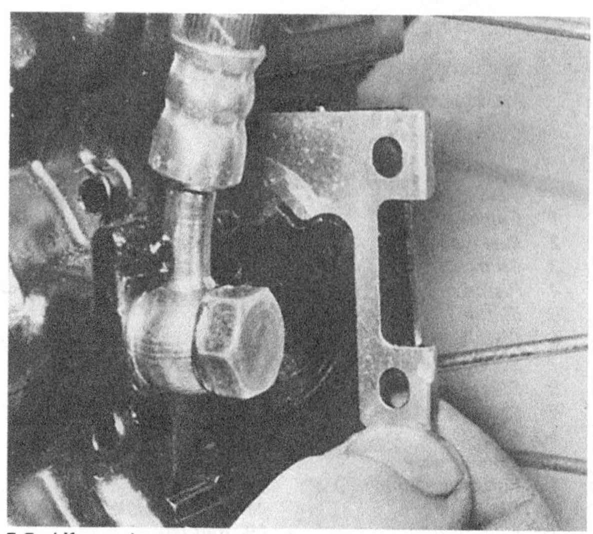
5.5a Lift out the outer pad, noting shim ...

5.5b ... followed by the inner pad

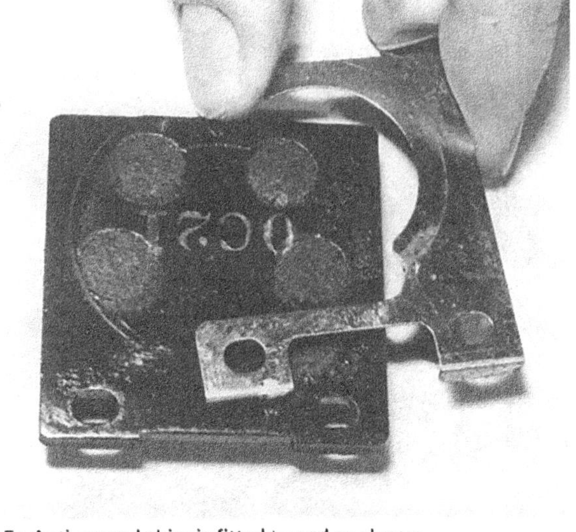

5.5c Anti-squeal shim is fitted to pad as shown

6.2 Disconnect brake pipe union and drain system

6.4a Remove bracket bolts and lift caliper clear

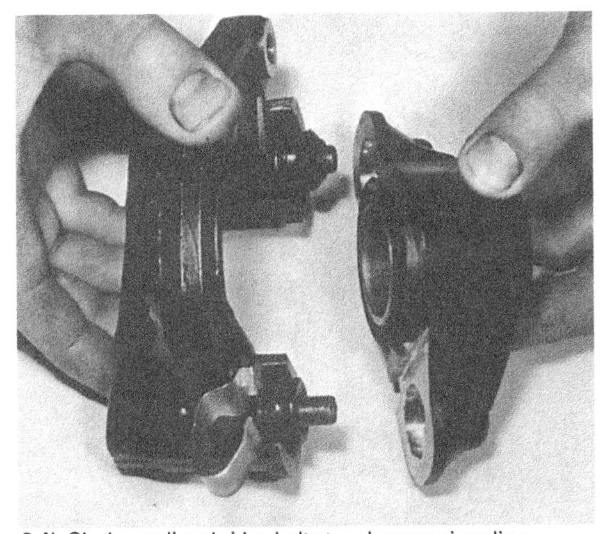

6.4b Slacken caliper bridge bolts to release main caliper

6.4c Bracket can be removed as shown

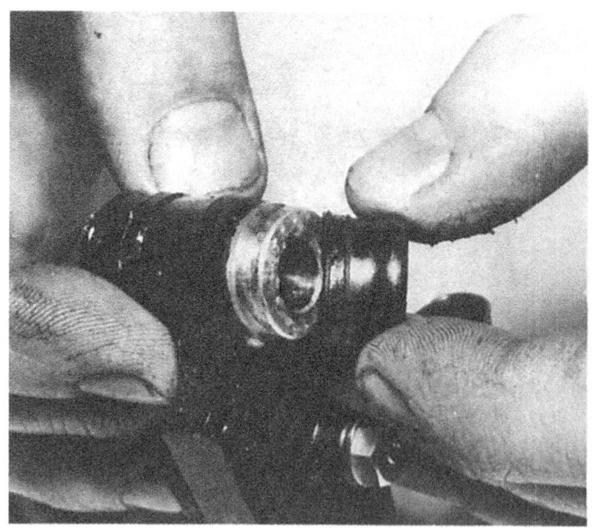
6.4d Note dust seals on bracket bosses

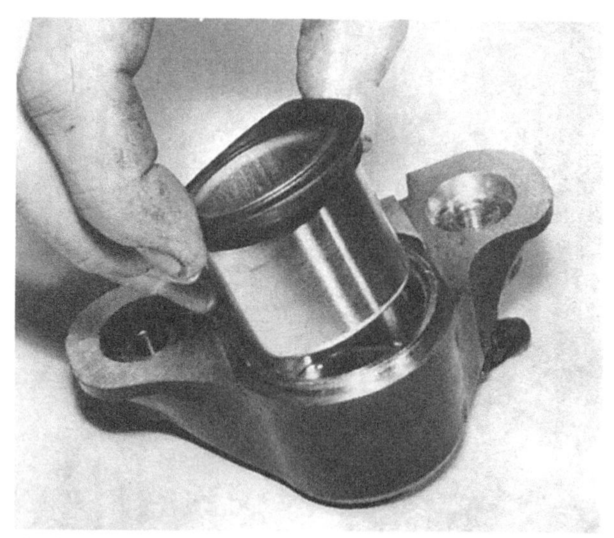
6.5 Displace piston using compressed air

6.7a Piston seal locates in groove around outer edge

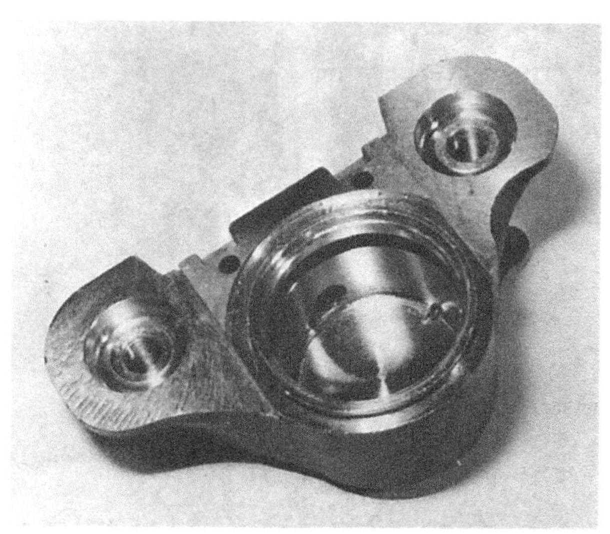
6.7b Check caliper bore for scores and check caliper seal

6.8a Ensure that piston (dust) seal locates correctly

6.8b Note sealing band on lower bridge bolt

7 Removing and replacing the brake disc

1 It is unlikely that the disc will require attention until a considerable mileage has been covered, unless premature scoring of the disc has taken place thereby reducing braking efficiency. To remove the disc, first detach the front wheel as described in Section 3 of this Chapter. The disc is bolted to the front wheel on the left-hand side by five bolts which pass into the hub.
2 The brake disc can be checked for wear and for warpage whilst the front wheel is still in the machine. Using a micrometer, measure the thickness of the disc at the point of greatest wear. If the measurement is much less than the recommended service limit of 4.0 mm (0.16 in) the disc should be renewed. Check the warpage of the disc by setting up a suitable pointer close to the outer periphery of the disc and spinning the front wheel slowly. If the total warpage is more than 0.3 mm (0.012 in) the disc should be renewed. A warped disc, apart from reducing the braking efficiency, is likely to cause juddering during braking and will also cause the brake to bind when it is not in use.

8 Master cylinder: examination and renovation

1 The master cylinder is unlikely to give trouble unless the machine has been stored for a long period or until a considerable mileage has been covered. The usual signs of trouble are leakage of fluid, causing a gradual fall in the fluid level and bad braking performance.
2 To gain access to the master cylinder, commence the dismantling operation by attaching a bleed tube to the caliper unit bleed valve. Open the bleed valve one complete turn, then operate the front brake lever until all the hydraulic fluid is pumped out of the reservoir. Close the bleed valve, remove the bleed tube and store the fluid in a closed container for subsequent reuse.
3 Remove the banjo bolt and fluid hose from the end of the master cylinder unit and remove the lever pivot bolt, the lever and the front brake switch. Detach the master cylinder from the handlebars by removing the two bolts and the securing clamp.
4 Access is now available to the piston and the cylinder and it is possible to remove the piston assembly, together with the relevant seals. Remove the circlip and sealing boot, followed by the next internal clip. The remainder of the components can be pushed out. Take note of the way in which the seals are fitted because they must be replaced in the same order and position. Failure to observe this necessity will result in brake failure.
5 Clean the master cylinder and piston with either hydraulic fluid or alcohol. On no account use abrasives or other solvents such as petrol. If any signs of damage or wear are evident, renewal is necessary. It is not practicable to reclaim either the piston or the cylinder bore.
6 Soak the new seals in hydraulic fluid for about 15 minutes prior to fitting, then reassemble the parts in exactly the same order, using the reversal of the dismantling procedure. Lubricate with hydraulic fluid and make sure the feathered edges of the seals are not damaged.
7 Refit the assembled master cylinder to the handlebars and reconnect the handlebar lever and hose. Refill the reservoir with hydraulic fluid and bleed the entire system by following the procedure described in Section 9 of this Chapter.
8 Check that the brake is working correctly before taking the machine on the road, to restore pressure and align the pads correctly. Use the brake gently for the first 50 miles or so to let the new components bed down correctly.
9 It should be emphasised that repairs to the master cylinder are best entrusted to a Honda Service Agent, or alternatively, that the defective parts should be replaced by a new unit. Dismantling and reassembly requires a certain amount of skill and it is imperative that the entire operation is carried out under surgically clean conditions.

9 Front disc brake: bleeding the hydraulic system

1 Removal of all the air from the hydraulic system is essential for the efficiency of the braking system. Air can enter the system due to leaks or when any part of the system has been dismantled for repair or overhaul. Topping the system up will not suffice, as air pockets will still remain, even small amounts causing dramatic loss of brake pressure.
2 Check the level in the reservoir, and fill almost to the top. Again, beware of spilling the fluid on to painted or plastic surfaces.
3 Place a clean jar below the brake caliper unit and attach a clear plastic tube from the caliper bleed screw to the container. Place some clean hydraulic fluid in the container so that the pipe is always immersed below the surface of the fluid.
4 Unscrew the bleed screw one complete turn and pump the handlebars lever slowly. As the fluid is ejected from the bleed screw the level in the reservoir will fall. Take care that the level does not drop too low whilst the operation continues, otherwise air will re-enter the system, necessitating a fresh start.
5 Continue the pumping action with the lever until no further air bubbles emerge from the end of the plastic pipe. Hold the brake lever against the handlebars and tighten the caliper bleed screw. Remove the plastic tube after the bleed screw is closed.
6 Check the brake action for sponginess, which usually denotes there is still air in the system. If the action is spongy, continue the bleeding operation in the same manner, until all traces of air are removed.
7 When all traces of air have been removed from the system, top up the reservoir and refit the diaphragm and cap or cover, as appropriate. Check the entire system for leaks, and check also that the brake system in general is functioning efficiently before using the machine on the road.
8 Brake fluid drained from the system will almost certainly be contaminated, either by foreign matter or more commonly by the absorption of water from the air. All hydraulic fluids are to some degree hygroscopic, that is, they are capable of drawing water from the atmosphere, and thereby degrading their specifications. In view of this, and the relative cheapness of the fluid, old fluid should always be discarded.

10 Front wheel bearings: examination and replacement

1 Place the machine on its centre stand and remove the front wheel as described in Section 3 of this Chapter. Remove the speedometer drive gearbox, then carefully prise out the oil seal taking care not to damage the sealing lip or the spring beneath it. An improvised tool can be made by heating the end of an old screwdriver and bending it into a hooked shape.
2 The wheel bearings can now be tapped out from each side with the use of a suitable long drift. Careful and even tapping will prevent the bearing 'tying' and damage to the races.
3 Remove all the old grease from the hub and bearings, giving the latter a final wash in petrol. When the bearings are clean lubricate them sparingly with a very light oil. Check the bearings for play and roughness when they are spun by hand. All used bearings will emit a small amount of noise when spun but they should not chatter or sound rough. If there is any doubt about the conditions of the bearings they should be renewed.
4 Before replacing the bearings pack them with high melting point grease. Do not overfill the hub centre with grease as it will expand when hot and may find its way past the oil seals. The hub space should be about $\frac{2}{3}$ full of grease. Drift the bearings in, using a soft drive on the outside ring of the bearing. Do not drive the centre ring of the bearing or damage will be incurred. Replace the oil seals carefully, drifting them into place with a thick walled tube of approximately the same dimension as the oil seal. A large socket spanner is ideal.

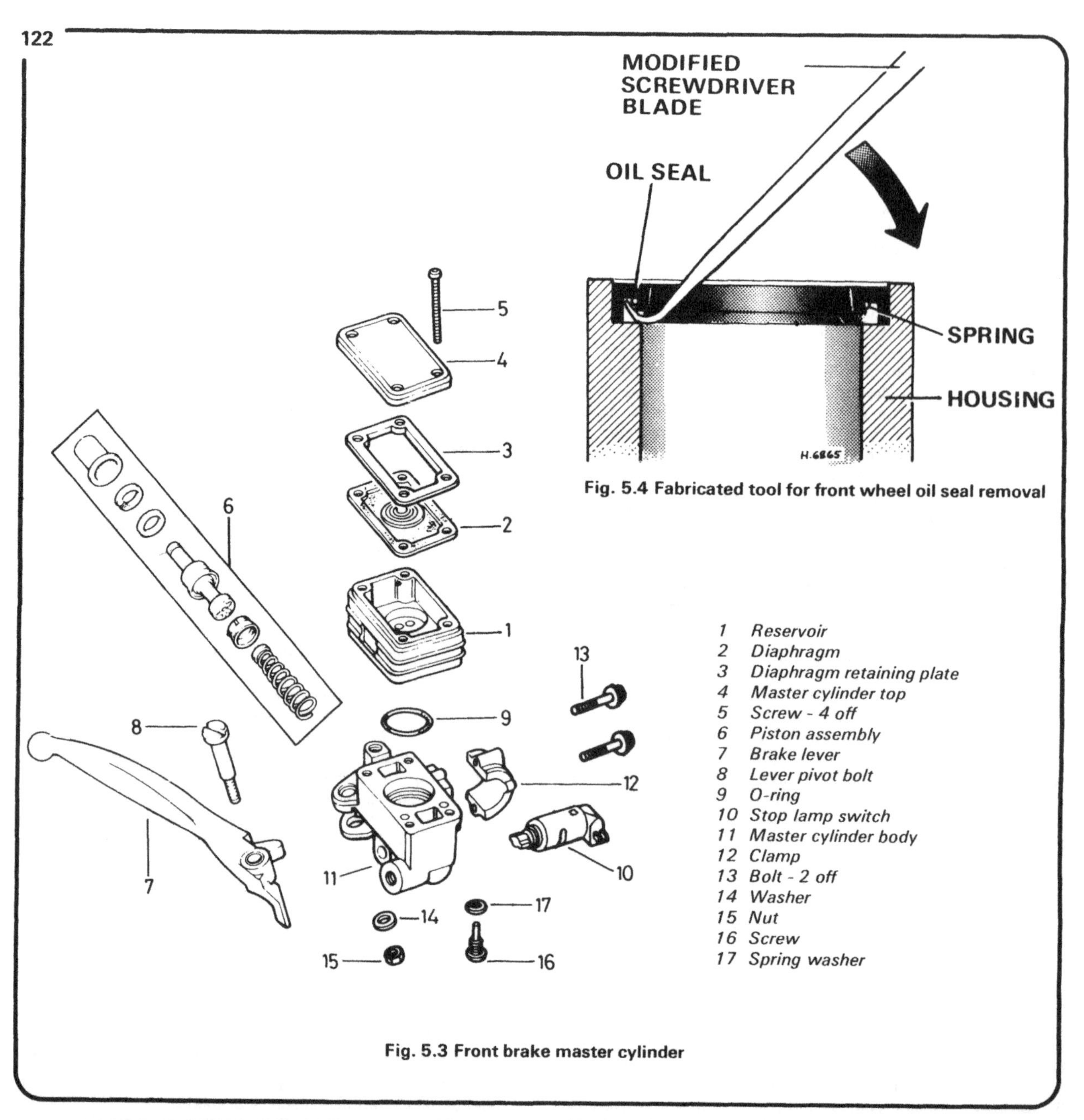

Fig. 5.4 Fabricated tool for front wheel oil seal removal

Fig. 5.3 Front brake master cylinder

9.3 Hydraulic system must be bled to remcve any air

10.1a Lift away the speedometer drive gearbox ...

10.1b ... and gently prise out the dust seal

10.1c Driving dog can now be displaced

10.1d Remove spacer from opposite side of hub ...

10.1e ... and remove dust seal to expose bearing

10.2a Use a long drift to remove bearings

10.2b Note spacer fitted between the bearings

10.4a Sealed face of bearings faces outwards

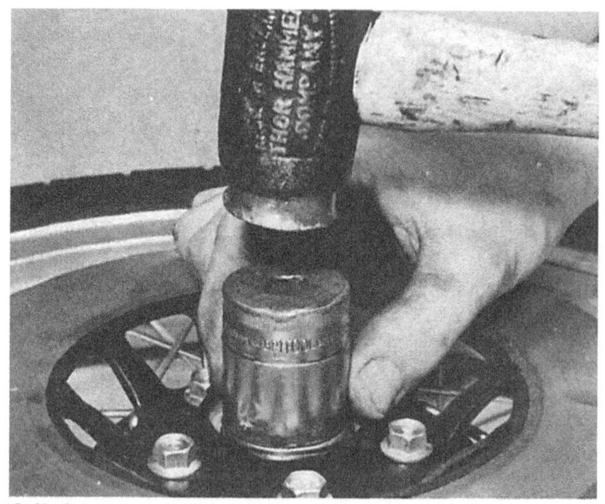

10.4b A large socket is useful when fitting bearings

11 Rear wheel: examination, removal and renovation

1 Place the machine on the centre stand so that the rear wheel is raised clear of the ground. Check for rim alignment, damage to the rim and loose or broken spokes by following the procedure relating to the front wheel, as described in Section 2 of this Chapter.

12 Rear wheel: removal and replacement

1 Place the machine on its centre stand so that the rear wheel is raised clear of the ground. Slacken and remove the rear brake adjuster nut to free the brake operating rod from the trunnion in the brake arm. Fit the trunnion and nut to the rod to avoid loss. Detach the brake torque arm from the back plate by removing the split pin and unscrewing the retaining nut.

2 Straighten and remove the split pin which secures the rear wheel spindle nut, and slacken the nut. Back off the chain tension adjusters so that the wheel can be pushed forward and the drive chain disengaged from the rear sprocket. Displace the end stops from the swinging arm fork ends. The wheel assembly can now be pulled back and clear of the fork.

3 Replace the wheel by reversing the dismantling procedure. The chain adjusters must not be omitted when inserting the wheel spindle. Before tightening the spindle nut fit the chain and check that the wheel is aligned correctly. See the instructions for chain adjustment in Section 18. Do not forget to secure the brake torque arm bolt by means of the split pin. If the torque arm becomes detached in service, the brake will lock on first application, causing an accident.

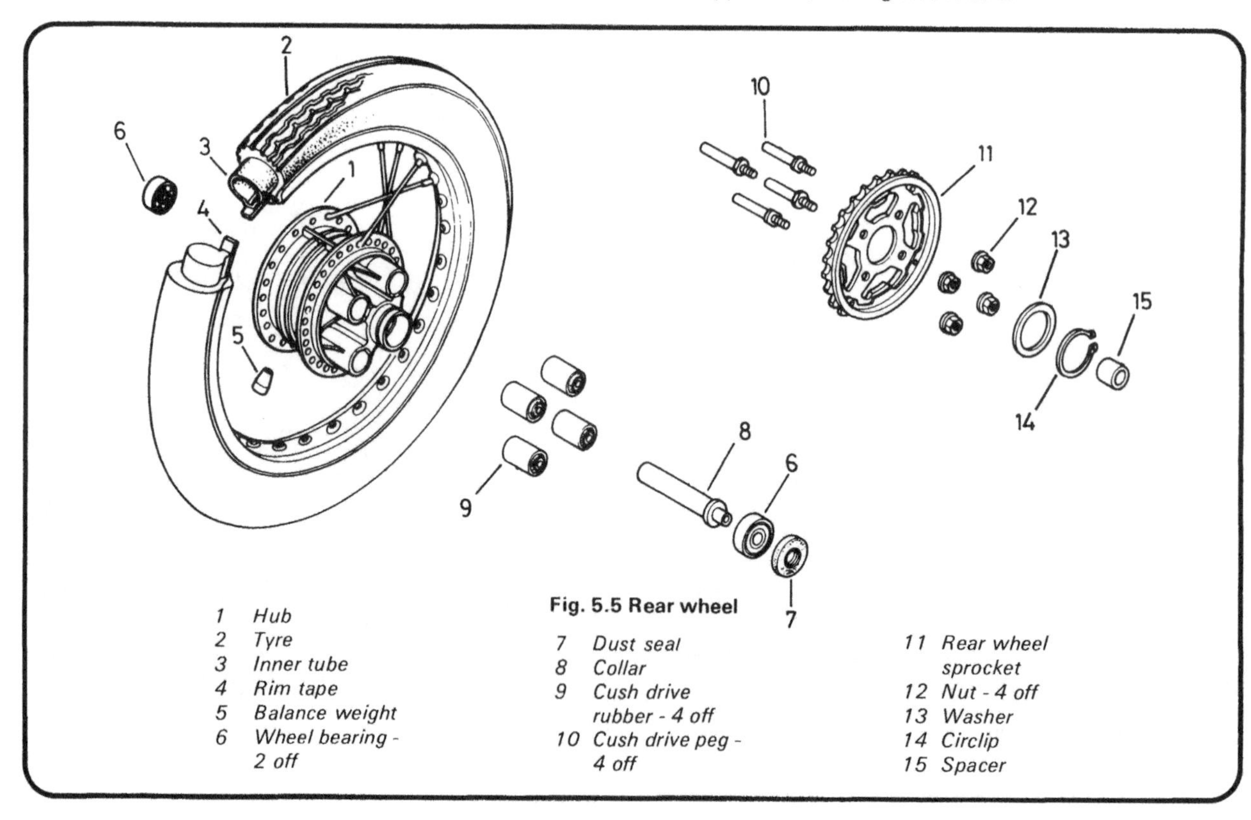

Fig. 5.5 Rear wheel

1 Hub
2 Tyre
3 Inner tube
4 Rim tape
5 Balance weight
6 Wheel bearing - 2 off
7 Dust seal
8 Collar
9 Cush drive rubber - 4 off
10 Cush drive peg - 4 off
11 Rear wheel sprocket
12 Nut - 4 off
13 Washer
14 Circlip
15 Spacer

12.1a Release brake rod, trunnion and spring from brake arm ...

12.1b ... and free the rear end of the torque arm

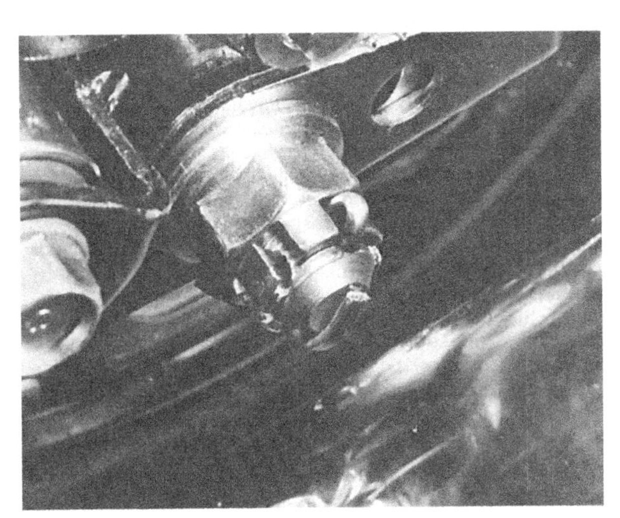
12.2a Remove split pin ...

12.2b ... and slacken spindle. Displace adjusters and fillets

12.2c Displace drive chain and pull wheel clear

13 Rear brake: examination and renovation

1 The rear brake is of the single leading shoe (sls) drum type, and is fitted with a brake wear indicator which provides a means of checking the condition of the linings. When the moving pointer reaches the index mark on the back plate, with the brake applied, the linings are in need of renewal. Access to the brake assembly is gained after the rear wheel has been removed as described in Section 12 of this Chapter.
2 When the wheel has been removed, the brake back plate assembly can be lifted away. Remove any accumulated dust from the brake drum by wiping it with a petrol-soaked rag. **Do not** blow the drum out with compressed air. The dust contains asbestos particles which are harmful if inhaled.
3 Measure the lining thickness at its thinnest point and renew the shoes if they are at or near the service limit shown in the Specifications Section. The shoes should also be renewed if they have become contaminated with oil or grease. The lining material is bonded to the shoes and this means that the shoes should be renewed complete. It is not practicable to re-line the shoes at home. When purchasing replacement shoes it is preferable to use genuine Honda parts. Many pattern types are available and whilst some may be of good quality, others may be dangerously sub-standard. Be particularly wary of pattern parts supplied with fake Honda packaging.

13.2 Remove brake back plate assembly and clean drum

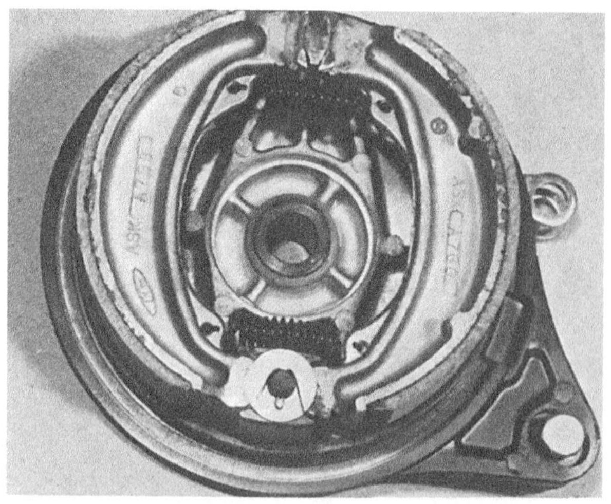

13.3a Check brake lining for wear and contamination

13.3b Drum should not exceed marked maximum diameter

14 Adjusting the rear brake

1 Adjustment of the rear brake is correct when there is 20 – 30 mm ($\frac{3}{4}$ – $1\frac{1}{4}$ in) up and down movement measured at the rear brake pedal foot piece, between the fully 'off' and 'on' position.

2 If, when the brake is fully applied, the angle between the brake arm and the operating rod is more than 90° the brake arm should be pulled off the camshaft, after loosening the pinch bolt. Reset the brake arm so that the right angle is produced.

3 The height of the brake pedal may be adjusted to suit the preference of the rider by means of a bolt and locknut fitted to a lug to the rear of the pedal. Note that it may be necessary to adjust the height-setting of the stop lamp switch after adjustment of the brake pedal position.

15 Rear wheel bearings: removal and replacement

1 To gain access to the rear wheel bearings, the wheel must be removed from the machine and the brake back plate withdrawn from the brake drum.

2 Knock out the bearings, using the same procedure as described for the front wheel, starting with the right-hand bearing and then the left-hand bearing. The remarks on bearing testing and regreasing described for the front wheel apply similarly to the rear wheel.

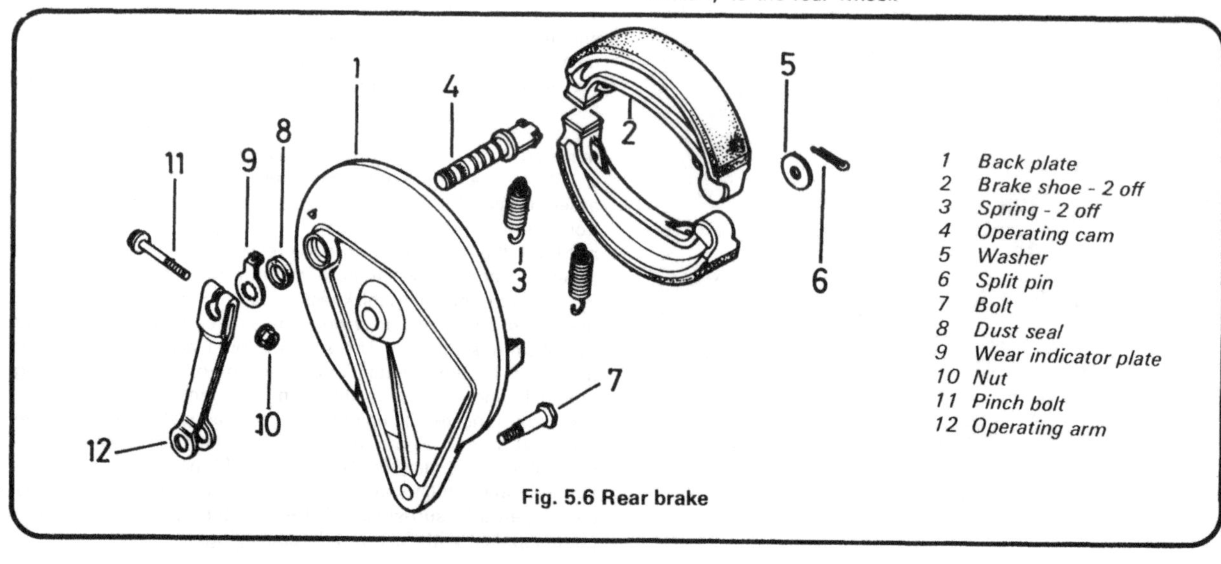

Fig. 5.6 Rear brake

Tyre changing sequence — tubed tyres

Deflate tyre. After pushing tyre beads away from rim flanges push tyre bead into well of rim at point opposite valve. Insert tyre lever adjacent to valve and work bead over edge of rim.

Use two levers to work bead over edge of rim. Note use of rim protectors.

Remove inner tube from tyre.

When first bead is clear, remove tyre as shown.

When fitting, partially inflate inner tube and insert in tyre.

Work first bead over rim and feed valve through hole in rim. Partially screw on retaining nut to hold valve in place.

Check that inner tube is positioned correctly and work second bead over rim using tyre levers. Start at a point opposite valve.

Work final area of bead over rim whilst pushing valve inwards to ensure that inner tube is not trapped.

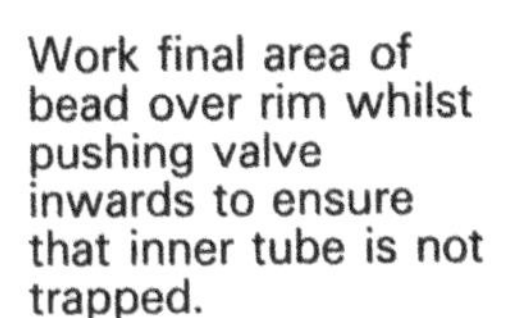

15.2a Remove the wheel spindle spacer ...

15.2b ... and lever out the oil seal

15.2c Bearings can be driven out in the same way ...

15.2d ... as described for the front wheel

16 Rear wheel sprocket: examination and replacement

1 The rear wheel sprocket is retained on the left-hand side of the hub by a large circlip, and is located by four pegs which pass into the cush drive in the hub and are retained by hexagonal nuts.
2 To remove the sprocket, detach the large circlip and the spacing plate which lies below. Pull the sprocket off the hub, together with the four drive pins. A cush drive is provided by four bonded rubber bushes pushed into bores in the hub. Where this method is employed it is probable that the sprocket pegs are seized in the steel sleeve bonded to the flexible rubbers. If this is the case, remove the peg securing nuts. The sprocket can then be lifted off with ease.
3 Check the condition of the sprocket teeth. If they are hooked, chipped or badly worn, the sprocket must be renewed. It is considered bad practice to renew one sprocket on its own. The final drive sprockets should always be renewed as a pair and a new chain fitted, otherwise rapid wear will necessitate even earlier renewal on the next occasion.
4 The sprocket may be refitted by reversing the dismantling procedure. It is important that the recesses in the rear of the sprocket are engaged correctly by the milled flats on each cush drive pin.

17 Rear wheel cush drive: exmaination and renovation

1 All models are fitted with a cush drive arrangement in the rear hub which allows a small amount of movement between the hub and the drive sprocket to help damp out transmission shock loads. The assembly consists of four tubular rubber bushes located in the hub. The four special pegs retained by nuts on the sprocket locate with these bushes, to give a cushioning effect to the sprocket and drive. To obtain access to the bushes, the sprocket has to be removed by detaching its circlip and pulling it from the wheel hub. Removal of the bushes is required when there is excessive sprocket movement. As stated in the previous Section, the pegs may seize after a considerable length of time. If this occurs, the sprocket complete with pegs should be drawn from the hub using a sprocket puller. A blanking plate or bar, fabricated from mild steel, will have to be made and placed over the bearing. The sprocket puller screw can then bear against the bar.
2 Removal of the flexible bushes is almost impossible without the use of a special expanding extractor. It is recommended that the wheel be returned to a Honda Service Agent who can carry out the work without risk of damage to the wheel.

16.2a Sprocket assembly is retained by large circlip

16.2b Lift assembly clear to expose cush drive rubbers

18 Chain: examination, lubrication and adjustment

1 The final drive chain is fully exposed apart from the protection given by a short chainguard along the upper run, and if not properly maintained will have a short life. A worn chain will cause rapid wear of the sprockets and they too will need replacement.
2 The chain tension will need adjustment at regular intervals, to compensate for wear. This is accomplished by loosening the rear wheel nut, which is secured by a split pin, and loosening the large nut holding the cush drive/sprocket hub hollow spindle, with the machine on the centre stand. The brake torque arm nuts should be loosened, but it is not necessary to remove the spring security pins.
3 Slacken the locknuts on the chain adjusters on the fork ends. Screw the adjusters inwards an equal amount to tighten the chain. The tension is correct if there is 15 – 25 mm ($\frac{3}{4}$ in – 1 in) up and down movement in the centre of the lower chain run. Always check the chain when it is at its tightest point; a chain rarely wears evenly. This may be accomplished by turning the wheel whilst applying a finger to the lower chain run. The tightest point is easily found.
4 Always adjust the draw bolts an even amount so that correct wheel alignment is preserved. The fork ends are marked with a series of vertical lines to provide a visual check. If desired, wheel alignment can be checked by running a plank of wood parallel to the machine so that it touches both walls of the rear tyre. If wheel alignment is correct, it should be equidistant from either side of the front wheel tyre when tested on both sides of the rear wheel; it will not touch the front tyre because this tyre has a smaller cross section. See the accompanying diagram.
5 Do not run the chain overtight to compensate for uneven wear. A tight chain will place excessive stresses on the gearbox and rear wheel bearings leading to their early failure. It will also absorb a surprising amount of power.
6 After a period of running the chain will require lubrication. Lack of oil will accelerate the rate of wear of both chain and sprockets and will lead to harsh transmission. The application of engine oil will act as a temporary expedient, but it is preferable to use one of the aerosol chain lubricants which are flung off the chain less readily than conventional oils. The endless construction of the chain means that all cleaning and lubrication must be carried out with the chain in position if extensive dismantling is to be avoided. To remove the chain from the machine it is first necessary to remove the rear wheel, gearbox sprocket and cover and the swinging arm assembly as described in Section 14 of Chapter 4.
7 The chain should be renewed as soon as it reaches the furthest limit of adjustment or if it is obviously loose and harsh in use. If renewal is necessary, check the condition of the gearbox and rear wheel sprockets. If these show signs of wear or damage, they must be renewed together with the chain. Worn sprockets will rapidly ruin a new chain.
8 When purchasing a new chain consideration should be given to fitting one with a joining link. There is no obvious reason for using an endless chain on a machine of this capacity and power output, and much dismantling work will be avoided when the time comes for future renewals. A chain with a joining link can be removed for more effective cleaning and lubrication. If this type of chain is chosen, it is important that the clip is fitted with the closed end facing in the direction of chain travel.
9 Replacement chains are now available in standard metric sizes from Renold Limited, the British chain manufacturer. When ordering a new chain, always quote the size, the number of chain links and the type of machine to which the chain is to

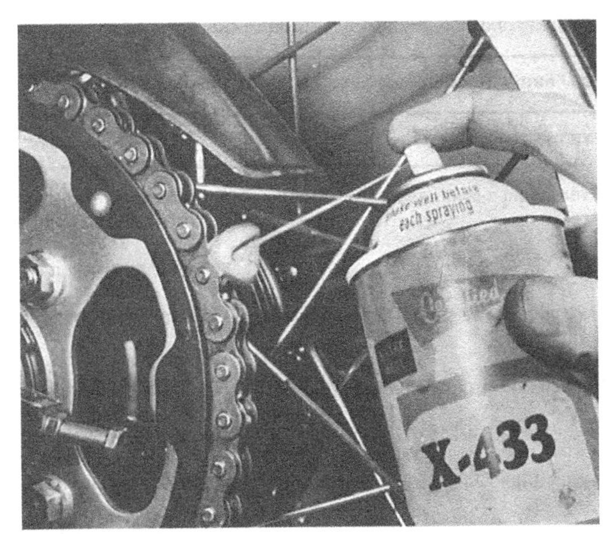

18.6 Original endless chain can be lubricated with aerosol

19 Tyres: removal and replacement

1 At some time or other the need will arise to remove and replace the tyres, either as the result of a puncture or because a renewal is required to offset wear. To the inexperienced tyre

changing represents a formidable task yet if a few simple rules are observed and the technique learned, the whole operation is surprisingly simple.

2 To remove the tyre from either wheel, first detach the wheel from the machine by following the procedure in Section 3 or 12 of this Chapter, depending on whether the front or the rear wheel is involved. Deflate the tyre by removing the valve insert and when it is fully deflated, push the bead of the tyre away from the wheel rim on both sides so that the bead enters the centre well of the rim. Remove the locking cap and push the tyre valve into the tyre itself.

3 Insert a tyre lever close to the valve and lever the edge of the tyre over the outside of the wheel rim. Very little force should be necessary; if resistance is encountered it is probably due to the fact that the tyre beads have not entered the well of the wheel rim all the way round the tyre.

4 Once the tyre has been edged over the wheel rim, it is easy to work around the wheel rim so that the tyre is completely free on one side. At this stage, the inner tube can be removed.

5 Working from the other side of the wheel, ease the other edge of the tyre over the outside of the wheel rim which is furthest away. Continue to work around the rim until the tyre is free completely from the rim.

6 If a puncture has necessitated the removal of the tyre, reinflate the inner tube and immerse it in a bowl of water to trace the source of the leak. Mark its position and deflate the tube. Dry the tube and clean the area around the puncture with a petrol soaked rag. When the surface has dried, apply the rubber solution and allow this to dry before removing the backing from the patch and applying the patch to the surface.

7 It is best to use a patch of the self-vulcanising type which will form a very permanent repair. Note that it may be necessary to remove a protective covering from the top surface of the patch, after it has sealed in position. Inner tubes made from synthetic rubber may require a special type of patch and adhesive if a satisfactory bond is to be achieved.

8 Before refitting the tyre, check the inside to make sure that the agent which caused the puncture is not trapped. Check the outside of the tyre, particularly the tread area, to make sure nothing is trapped that may cause a further puncture.

9 If the inner tube has been patched on a number of past occasions, or if there is a tear or large hole, it is preferable to discard it and fit a new one. Sudden deflation may cause an accident, particularly if it occurs with the front wheel.

10 To replace the tyre, inflate the inner tube sufficiently for it to assume a circular shape but only just. Then push it into the tyre so that it is enclosed completely. Lay the tyre on the wheel at an angle and insert the valve through the rim tape and the hole in the wheel rim. Attach the locking cap on the first few threads, sufficient to hold the valve captive in its correct location.

11 Starting at the point furthest from the valve, push the tyre bead over the edge of the wheel rim until it is located in the central well. Continue to work around the tyre in this fashion until the whole of one side of the tyre is on the rim. It may be necessary to use a tyre lever during the final stages.

12 Make sure that there is no pull on the tyre valve and again commencing with the area furthest from the valve, ease the other bead of the tyre over the edge of the rim. Finish with the area close to the valve, pushing the valve up into the tyre until the locking cap reaches the rim. This will ensure the inner tube is not trapped when the last section of the bead is edged over the rim with a tyre lever.

13 Check that the inner tube is not trapped at any point. Reinflate the inner tube and check that the tyre is seating correctly around the wheel rim. There should be a thin rib moulded around the wall of the tyre on both sides which should be equidistant from the wheel rim at all points. If the tyre is unevenly located on the rim, try bouncing the wheel when the tyre is at the recommended pressure. It is probable that one of the beads has not pulled clear of the centre well.

14 Always run the tyres at the recommended pressures and never under or over-inflate. The correct pressures for solo use are given in the Specifications Section of this Chapter. If a pillion passenger is carried, increase the rear tyre pressure only by approximately 4 psi.

15 Tyre replacement is aided by dusting the side walls, particularly in the vicinity of the beads, with a liberal coating of French chalk. Washing up liquid can also be used to good effect but this has the disadvantage of causing the inner surfaces of the wheel rim to corrode.

16 Never replace the inner tube and tyre without the rim tape in position. If this precaution is overlooked there is a good chance of the ends of the spoke nipples chafing the inner tube and causing a crop of punctures.

17 Never fit a tyre which has a damaged tread or side walls. Apart from the legal aspects, there is a very great risk of a blow-out, which can have serious consequences.

20 Fault diagnosis: wheels, brakes and tyres

Symptom	Cause	Remedy
Handlebars oscillate at low speeds	Buckle or flat in wheel rim, most probably front wheel	Check rim alignment by spinning wheel. Correct by retensioning spokes or rebuilding on new rim.
	Tyre not straight on rim	Check tyre alignment.
Machine lacks power and accelerates poorly	Rear brake binding	Warm brake drum provides best evidence Re-adjust brake.
Rear brake grabs when applied gently	Ends of brake shoes not chamfered	Chamfer with file.
	Elliptical brake drum	Lightly skim in lathe (specialist attention required).
Front brake feels spongy	Air in hydraulic system	Bleed brake.
Brake pull-off sluggish	Brake cam binding in housing (drum brake only)	Free and grease.
	Weak brake shoe springs (drum brake only)	Renew if springs have not become displaced.
	Sticking pistons in brake caliper	Overhaul caliper unit.
Harsh transmission	Worn or badly adjusted final drive chain	Adjust or renew as necessary
	Hooked or badly worn sprockets	Renew as a pair.
	Worn or deteriorating cush drive rubbers	Renew rubbers.

Chapter 6 Electrical system

For information relating to the CB250 RSD-C model, refer to Chapter 7

Contents

Specifications

Battery

Make	Yuasa
Capacity	12v 9Ah
Earth connection	Negative (–)

Alternator

Charge starts at	1500 rpm
Charging output:	
Minimum	9.5A @ 5000 rpm
Maximum	15.0A @ 8000 rpm

Regulator/rectifier

Type	Combined transistorised unit

Fuses

Headlamp circuit	7 amp
Tail lamp circuit	7 amp
Main	15 amp

Bulbs

Headlamp	12V 45/40W
Stop/tail	12V 5/21W
Indicators	12V 21W
Parking lamp	12V 4W
Instrument illumination	12V 3.4W (2 off)
Warning lamps	12V 3.4W (3 off)

1 General description

All models covered by this manual are fitted with a 12 volt electrical system. The circuit comprises a crankshaft mounted permanent magnet alternator and a combined solid-state regulator rectifier unit. The regulator maintains the output to within a specified limit to prevent overcharging and the rectifier converts the ac (alternating current) output to dc (direct current) to enable the lights and ancillary equipment to be powered and to allow the battery to be charged. The alternator consists of a multi-coil stator bolted to the left-hand casing and a permanent magnet rotor mounted on the crankshaft end.

2 Crankshaft alternator: checking the output

1 The output from the alternator mounted on the end of the crankshaft can be checked only with special test equipment of the multimeter type. It is unlikely that the average owner/rider will have access to this equipment or instruction in its use. In consequence, if the performance of the alternator is in any way suspect, it should be checked by a Honda Service Agent or an auto-electrical specialist.

2 If a multimeter is available, a general check on the alternator may be carried out as follows. Connect a dc voltmeter across the battery by removing the battery positive (+) lead and connecting one of the voltmeter probes to the positive (+) battery terminal. The remaining probe lead should be connected to a convenient earth point. An ammeter should now be connected between the battery positive (+) terminal and the battery positive (+) lead. Disconnect the regulator/rectifier black wire for this test. The battery must be in good condition and fully charged, and the engine must be warmed up to normal running temperature before the test. Note that for certain parts of the test the engine must be run at high speed; ensure this is kept to a minimum to prevent engine damage.

3 Start the engine and switch the headlamp onto main beam. The ammeter readings should be as follows at the specified engine speeds:

1500 rpm	initial charge starts
5000 rpm	9.0 A minimum
8000 rpm	15.0 A maximum

Actual voltage readings are not specified by the manufacturer although approximately 14 – 15 volts should be indicated.

4 If the voltage and current is correct, it may be assumed that the system is functioning properly. Do not forget to reconnect the regulator/rectifier black wire when the test is complete. A marked reduction in output may be a result of damaged windings in the stator coils or broken leads. These may be checked for continuity and resistance without removing the alternator from the machine, as follows.

5 Remove the seat and fuel tank to gain access to the alternator and rectifier/regulator unit wiring. Two 3-pin connectors are involved, one carrying the three yellow alternator output leads and the other the red/white, green and black leads from the rectifier. Stator continuity is checked between each of the three yellow wires.

6 Place the multimeter in the resistance function and check for continuity between the yellow leads. Check also continuity between each lead and an earth point on the engine. If no continuity exists between any two leads or if continuity exists between any lead and an earth point, renewal of the alternator is almost certained required. Lack of continuity may be due to a broken wire which can, in some cases, be repaired.

7 If the resistance tests are satisfactory, the regulator/rectifier unit should be examined as described in the following Section. This requires access to both of the above mentioned connectors, so the tank and seat should be left off at this stage.

1.1 Stator is bolted to inside of engine casing

3 Regulator/rectifier: testing

1 This component is a heavily-finned sealed metal unit bolted to the frame beneath the fuel tank. If the unit is found to be damaged or faulty, it must be renewed, even if only one side is affected; repairs are not possible.

2 To check the regulator side, a voltmeter must be connected across the battery terminals. With the engine running, the voltage across the battery should be between 14 – 15V.

3 To test the rectifier, separate the leads from the unit at the two block connectors. Using a multimeter set to the appropriate resistance scale, check for continuity between the green wire and each yellow wire and between the red/white wire and each yellow wire, then reverse the meter probes and check for continuity in the opposite direction. In the normal direction of current flow the meter should show very little resistance (approx 5 – 40 ohm) but in the reverse direction much greater resistance (approx 2 K ohms) should be measured. If any of the twelve tests does not produce the expected result then that particular diode is faulty and the complete unit must be renewed.

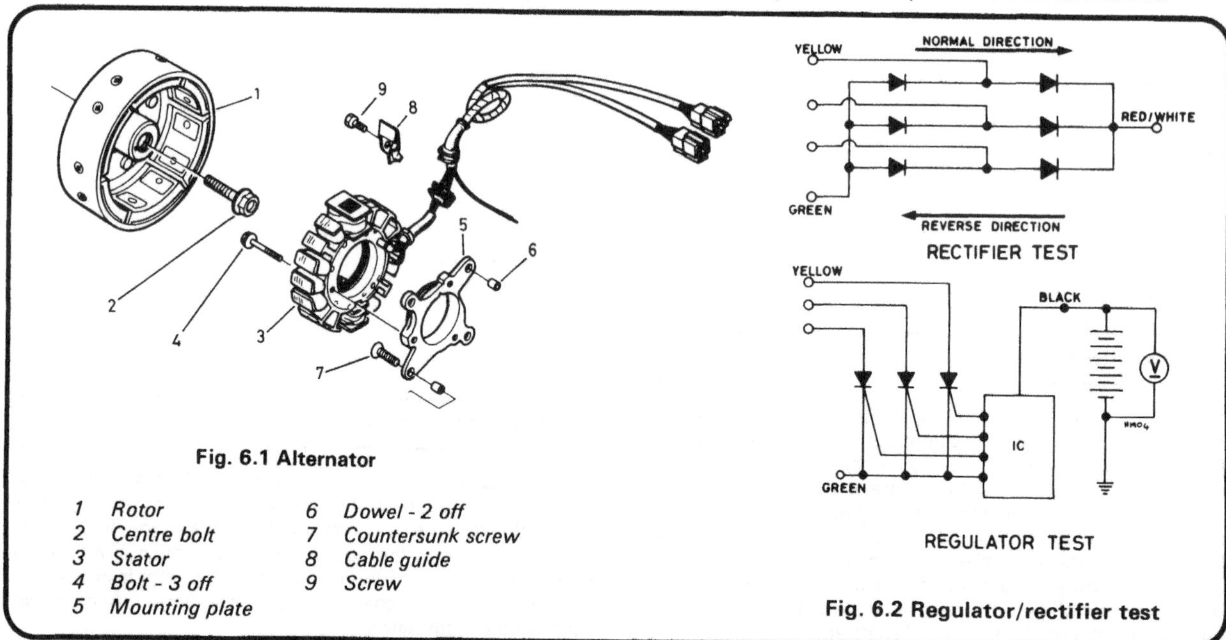

Fig. 6.1 Alternator

1 *Rotor*
2 *Centre bolt*
3 *Stator*
4 *Bolt - 3 off*
5 *Mounting plate*
6 *Dowel - 2 off*
7 *Countersunk screw*
8 *Cable guide*
9 *Screw*

REGULATOR TEST

Fig. 6.2 Regulator/rectifier test

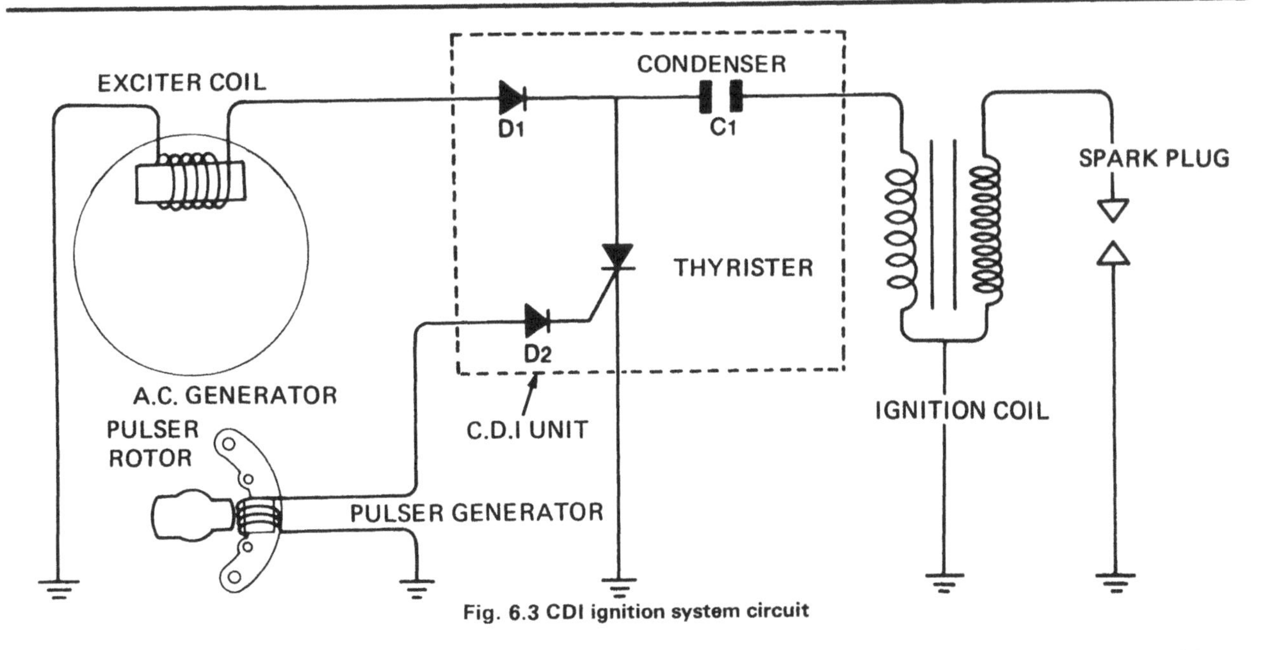

Fig. 6.3 CDI ignition system circuit

4 Battery: examination and maintenance

1 All models are fitted with a lead-acid battery of 12 volt, 9 Ampere-hour (12v 9Ah) capacity.

2 The transparent plastic case of the battery permits the upper and lower levels of the electrolyte to be observed when the battery is pulled from its housing below the dual seat. Access to the battery may be gained by detaching the frame right-hand side cover. Maintenance is normally limited to keeping the electrolyte level between the prescribed upper and lower limits and by making sure the vent pipe is not blocked. The lead plates and their separators can be seen through the transparent case, a further guide to the general condition of the battery.

3 Unless acid is spilt, as may occur if the machine falls over, the electrolyte should always be topped up with distilled water to restore the correct level. If acid is spilt on any of the machine, it should be neutralised with an alkali such as washing soda and washed away with plenty of water, otherwise serious corrosion will occur. Top up with sulphuric acid of the correct specific gravity (1.260 – 1.280) only when spillage has occurred. Check that the vent pipe is well clear of the frame tubes or any of the other cycle parts, for obvious reasons.

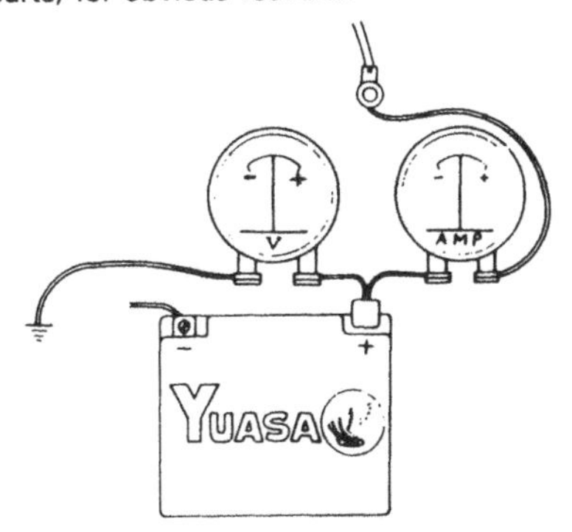

Fig. 6.4 Charging system output test connections

5 Battery: charging procedure

1 The normal charging rate for any battery is 1/10 the rated capacity. Hence the charging rate for the 9 Ah battery is 0.9 amps. A slightly higher rate of charge may be used in an emergency but this should not exceed 1.2A. The higher charge rate should, if possible, be avoided since it will shorten the working life of the battery.

2 Make sure that the battery charger connections are correct, red to positive and black to negative. It is preferable to remove the battery from the machine whilst it is being charged and to remove the vent plug from each cell. When the battery is reconnected to the machine, the black lead must be connected to the negative terminal and the red lead to positive. This is most important, as the machine has a negative earth system. If the terminals are inadvertently reversed, the electrical system will be damaged permanently.

6 Fuse: location and replacement

1 The electrical system is protected by a bank of three fuses, the headlamp and tail lamp circuits each having a 7 Amp fuse, whilst the main fuse is rated at 15 Amps. The fuses are housed in a plastic holder mounted next to the battery. The lid of the fuse box carries a spare fuse of both of the above values.

2 Before replacing a fuse that has blown, check that no obvious short circuit has occurred, otherwise the replacement fuse will blow immediately it is inserted. It is always wise to check the electrical circuit thoroughly, to trace the fault and eliminate it.

3 When a fuse blows while the machine is running and no spare is available, a 'get you home' remedy is to remove the blown fuse and wrap it in silver paper before replacing it in the fuse holder. The silver paper will restore the electrical continuity by bridging the broken fuse wire. This expedient should NEVER be used if there is evidence of short circuit or other major electrical fault, otherwise more serious damage will be caused. Replace the 'doctored' fuse at the earliest possible opportunity, to restore full circuit protection.

7 Headlamp: replacing the bulbs and adjusting beam height

1 In order to gain access to the headlamps bulbs it is necessary to first remove the rim, complete with the reflector and headlamp glass. The rim is retained by two screws which pass through the headlamp shell on the underside of the unit.

2 Pull the headlamp bulb socket from the rear bulb holder. The headlamp bulb is retained by a spring loaded collar. To

release the collar, depress and then twist it in an anti-clockwise direction. The collar, spring and bulb may be lifted from position. A reflector that accepts a pilot bulb is fitted to all models delivered to countries or states where parking lights are a statutory requirement. The pilot bolt is held in the bulb holder by a bayonet fixing.

3 Beam height on all models is effected by tilting the headlamp shell after the mounting bolts have been loosened slightly.

4 In the UK, regulations stipulate that the headlamps must be arranged so that the light will not dazzle a person standing at a distance greater than 25 feet from the lamp, whose eye level is not less than 3 feet 6 inches above that plane. It is easy to approximate this setting by placing the machine 25 feet away from a wall, on a level road, and setting the dipped beam height so that it is concentrated at the same height as the distance of the centre of the headlamp from the ground. The rider must be seated normally during this operation, and also the pillion passenger, if one is carried regularly.

8 Stop and tail lamp: replacement of bulbs

1 The combined stop and tail lamp bulb contains two filaments, one for the stop lamp and one for the tail lamp.

2 The offset pin bayonet fixing bulb can be removed after the plastic lens cover and screws have been removed.

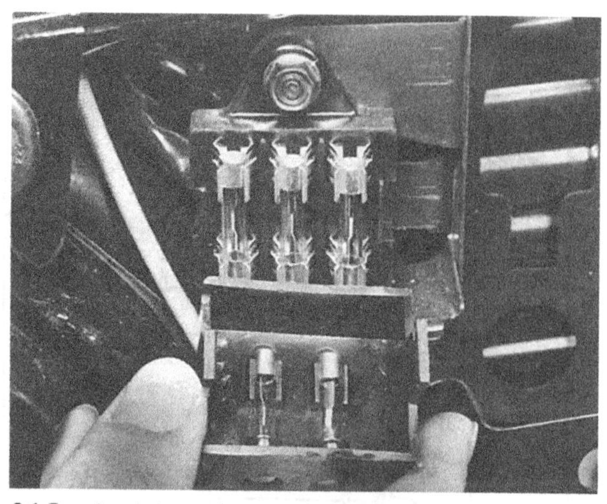

6.1 Fuse box is located next to the battery

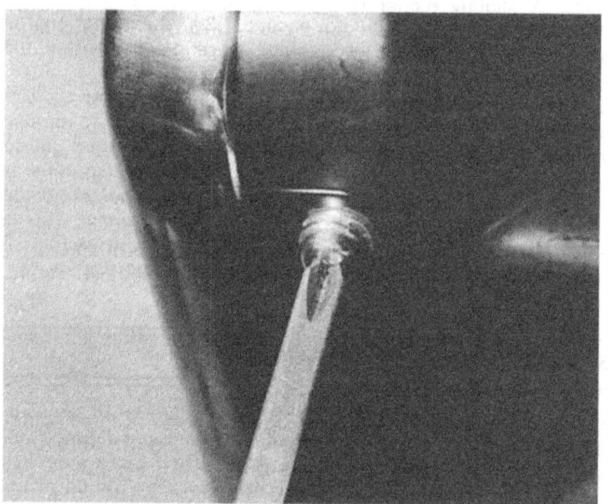

7.1 Release screws on underside of headlamp to free unit

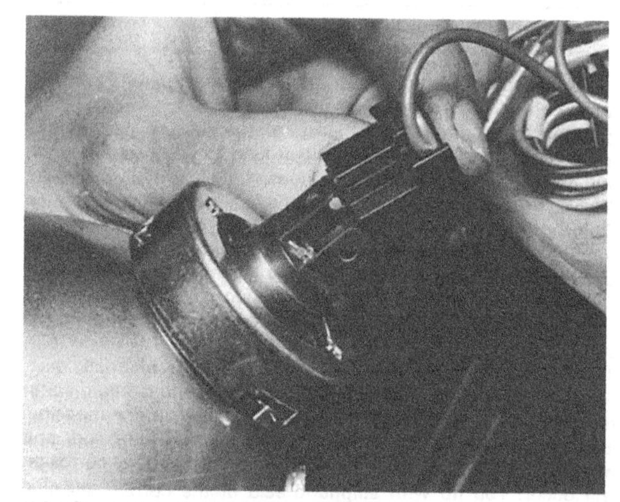

7.2a Connector plugs into rear of bulb

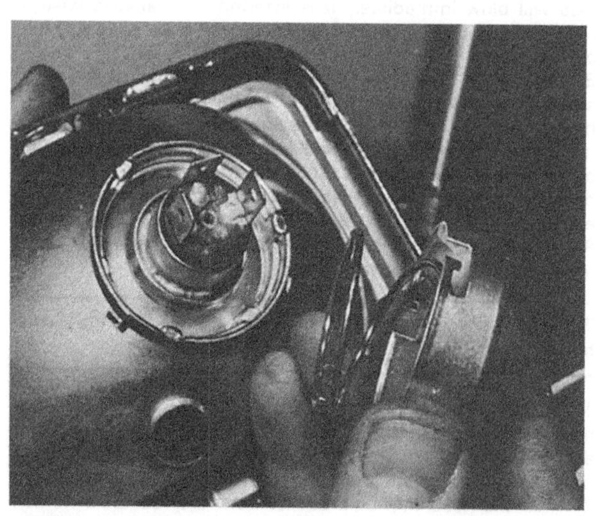

7.2b Twist retaining ring and remove it and the spring

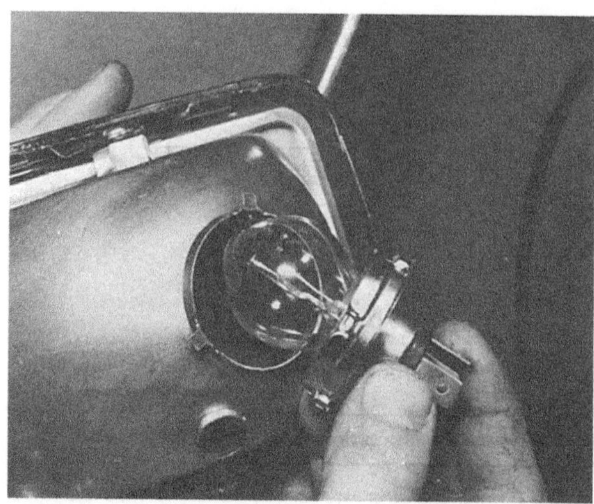

7.2c Bulb is located by three tangs around its flange

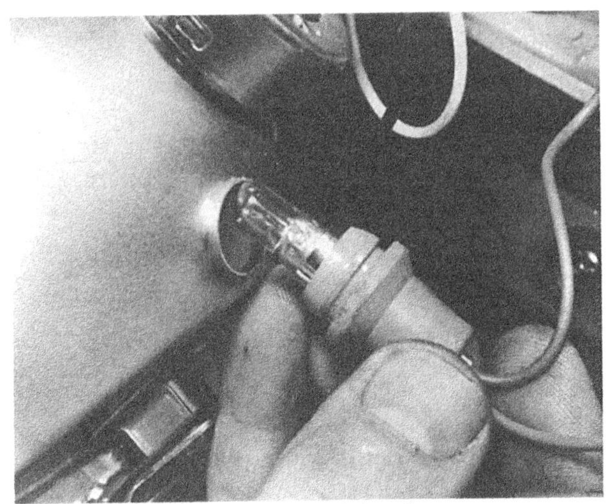
7.2d Parking bulb holder is a push fit in reflector

7.4 Screw in rim controls horizontal alignment

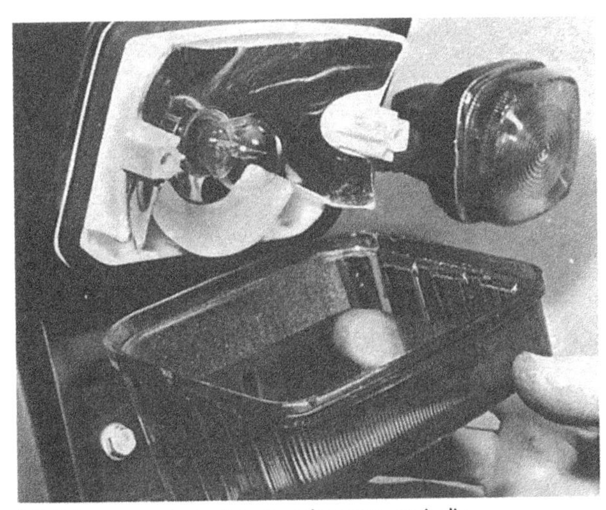
8.2 Release rear lamp lens to gain access to bulb

9 Flashing indicator lamps: replacing the bulb

1 The flashing indicators are mounted on resilient stalks designed to protect them from accidental knocks and the bulbs from vibration. If a bulb or wiring connection fails, the affected lamp will cease operation, the failure being indicated by rapid flashing of the remaining lamp and warning light.
2 The lens is clipped to the body of the lamp and can be removed by using a coin or broad-bladed screwdriver in the slot provided to lever the lens off. Both the lens and the body are of plastic construction, so care must be taken to avoid damage during removal The bulb is of the conventional bayonet type and can be removed by pushing inwards, twisting gently anti-clockwise and releasing.
3 No reflectors are fitted to the indicator lamps, and owners may wish to improve their visibility in bright sunlight by lining the inside of the lamp body with aluminium cooking foil. The lamps are, however, adequate in most normal conditions.

10 Flashing indicator relay: location and renewal

1 The indicator system is controlled by a flasher relay mounted to the front of the battery tray, beneath the right-hand side panel. The unit is retained by a rubber mounting which isolates it from road shocks and vibration.
2 When the relay malfunctions, it must be renewed; a repair is impracticable. When the unit is in working order audible clicks will be heard which coincide with the flash of the indicator lamps. If the lamps malfunction, check firstly that a bulb has not blown, or the handlebar switch is not faulty. The usual symptom of a fault is one initial flash before the unit goes dead.
3 Take great care when handling a flasher unit. It is easily damaged, if dropped.

11 Instrument panel and warning lamp bulbs: renewal

1 The various bulbs in the instrument panel are held in rubber holders which are a push fit in the underside of the panel. To gain access to the holders it is best to release the panel so that it can be tilted upwards. The panel is secured by two rubber-mounted retaining bolts which pass down into the top yoke. When these have been removed it is possible to lift and turn the assembly to expose the bulbholders, which can be pulled out of their recesses.
2 The bulbs are of the bayonet cap type and are released by depressing and twisting them anti-clockwise. When purchasing replacement bulbs ensure that they are of the specified voltage and wattage, 12v 3.4w.

12 Horn: location and adjustment

1 The horn is mounted on the underside of the steering head via a flexible mounting strip. No maintenance is required, other than regular cleaning to remove road dirt. If the horn malfunctions, check that power is reaching it by substituting a suitable 12 volt bulb and operating the horn button with the ignition switched on. If the bulb works, try adjusting the horn, using the adjuster screw and locknut on the back of the unit. It is not possible to dismantle a defective horn, renewal being the only practicable solution.

13 Ignition switch: removal and testing

1 The ignition, or main, switch is secured to the underside of the instrument panel by two bolts. The procedure for removing and testing it is described in Section 4 of Chapter 3.

9.2a Indicator lenses can be prised off as shown

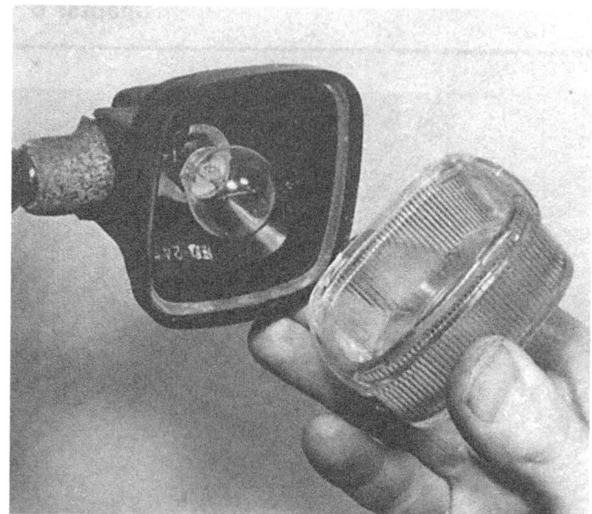

9.2b Bulb is a bayonet fitting

10.1 Flasher relay (unit) is mounted forward of battery

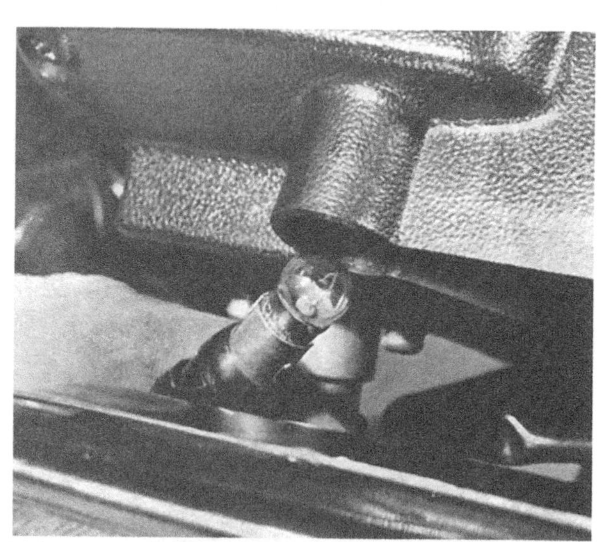

11.1 Instrument panel bulbs are a push fit in panel

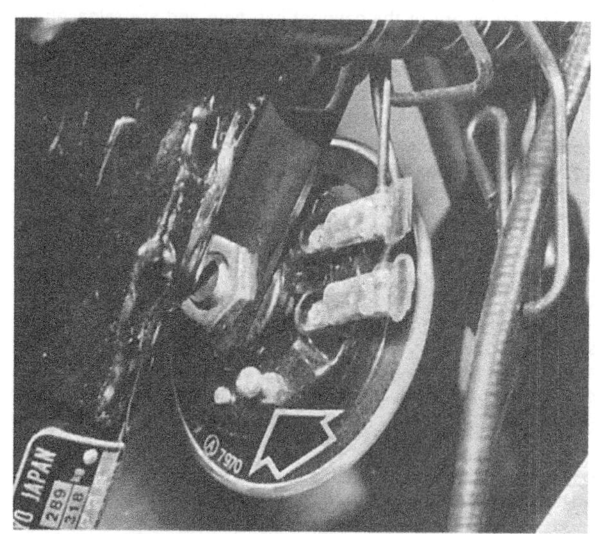

12.1 Horn location – note adjuster screw (arrowed)

13.1 Ignition switch is bolted to underside of panel

14 Stop lamp switch: adjustment

1 All models have a stop lamp switch fitted to operate in conjunction with the rear brake pedal. The switch is located immediately to the rear of the crankcase, on the right-hand side of the machine. It has a threaded body giving a range adjustment.
2 If the stop lamp is late in operating, turn the adjuster nut in a clockwise direction so that the switch rises from the bracket to which it is attached.
3 If the lamp operates too early, the adjuster nut should be turned anti-clockwise so that the switch body is lowered in relation to the mounting bracket.
4 As a guide, the light should operate after the brake pedal has been depressed by about 2 cm ($\frac{3}{4}$ inch).
5 A stop lamp switch is also incorporated in the front brake, to give warning when the front brake is applied. This is not yet a statutory requirement in the UK, although it applies in many other countries and states.
6 The front brake stop lamp switch is built into the hydraulic system and contains no provision for adjustment. If the switch malfunctions, it must be renewed.

15 Neutral indicator switch: location and testing

1 The neutral indicator switch comprises a metal contact blade mounted on the end of the gear selector drum and a fixed terminal mounted in the left-hand outer cover. The two coincide when neutral has been selected, allowing the lead from the warning lamp to earth, completing the circuit.
2 If the warning lamp fails to light when neutral is selected and the ignition is switched on, remove the gearbox sprocket cover to expose the fixed contact. Free the lead by pulling the plastic collar outwards. Earth the lead against a convenient unpainted area of the crankcase, select neutral and turn the ignition switch on. If the neutral indicator lamp does not light, check the bulb and wiring for failure. If it does light, the switch itself is at fault.
3 Remove the left-hand outer cover and check that the fixed terminal is in position and that the contact blade has not become flattened or broken. If the latter is removed for examination or renewal, ensure that it is refitted in the correct position. Unless care is taken it can be fitted 180° from the correct location.

16 Handlebar switches: general

1 Generally speaking. the switches give little trouble, but if necessary they can be dismantled by separating the halves which form a split clamp around the handlebars. Note that the machine cannot be started until the ignition cut-out on the right-hand end of the handlebars is turned to the central 'ON' position.
2 Always disconnect the battery before removing any of the switches, to prevent the possibility of short circuit. Most troubles are caused by dirty contacts, but in the event of the breakage of some internal part, it will be necessary to renew the complete switch.
3 Because the internal components of each switch are very small, and therefore difficult to dismantle and reassemble, it is suggested a special electrical contact cleaner be used to clean corroded contacts. This can be sprayed into each switch, without the need for dismantling.
4 If the switch unit is found to be faulty and the contact cleaner fails to effect a cure, it is worthwhile attempting to dismantle and physically clean the contacts because the switch will otherwise require renewal. If it becomes necessary to obtain a replacement unit, try a motorcycle breaker first. New switch assemblies are very expensive and a sound secondhand item can prove a far better alternative.

16.1a The right-hand handlebar switch assembly

16.1b The left-hand handlebar switch assembly

16.3 Separate switch halves and clean with aerosol cleaner

17 Fault diagnosis: electrical system

Symptom	Cause	Remedy
Complete electrical failure	Blown fuse	Check wiring and electrical components for short circuit before fitting a new fuse. Check battery connections, also whether connections show signs of corrosion.
Dim lights, horn inoperative	Discharged battery	Recharge battery with battery charger and check whether alternator is giving correct output (electrical specialist)
Constantly 'blowing' bulbs	Vibration, poor earth connection	Check whether bulb holders are secured correctly. Check earth return or connections to frame.

Left-hand view of the CB250 RSD-C model

Chapter 7 The CB250 RSD-C model

Contents

Specifications

Information is given only where it differs from that given in the Specifications Sections of Chapters 1 to 6

Model dimensions and weights

Overall length	2070 mm (81.5 in)
Overall width	730 mm (28.7 in)
Overall height	1060 mm (41.7 in)
Dry weight	131 kg (289 lb)

Specifications relating to Chapter 1

Engine

Compression ratio	9.5 : 1

Specifications relating to Chapter 2

Carburettor

Type	PD70E

Specifications relating to Chapter 3

Ignition timing

Initial	15° BTDC @ 1200 rpm (F mark aligned)
Advance starts	N/Av
Full advance	37 ± 2° BTDC @ 3450 rpm

Specifications relating to Chapter 5

Torque wrench settings

Sprocket retaining bolts	2.9 – 3.5 kgf m (21 – 25 lbf ft)

Specifications relating to Chapter 6

Alternator

Output @ 5000 rpm:	
CB250 RS-A	110W
CB250 RSD-C	170W

Starter motor

Brush length	13.0 mm (0.5118 in)
Service limit	6.5 mm (0.2559 in)
Brush spring pressure	400 ± 60 grams (14.11 ± 2.12 oz)

Bulbs

Headlamp	12V 60/55W

1 Introduction to the CB250 RSD-C

As the CB250 RS-A continued unchanged until phased out in March 1983 this Chapter covers the CB250 RSD-C model, describing only those features which require a different working procedure. When working on such a machine, refer first to this Chapter to note any changes in specification or working procedure, then refer to the relevant part of the main text.

The CB250 RSD-C model, also known as the CB250 RS-DX or as the CB250 RS Deluxe, was introduced in January 1982 first as a supplement to, and then replacing, the standard CB250 RS-A. The main difference between the two models was the addition of an electric starter, the kickstart and automatic decompressor being deleted to minimise the inevitable increase in weight. An O-ring final drive chain and modified rear wheel cush drive were fitted to counter excessive transmission snatch at low speeds which caused premature chain wear on the CB250 RS-A model. Apart from the other modifications noted in this Chapter, the styling of the new model was slightly altered and a lockable storage compartment fitted in the seat/tail unit.

Note: A new model, the CBX250 RS-E, was introduced in 1984. This features a new engine/gearbox unit of 249cc capacity with an RFVC-type cylinder head operated by double overhead camshafts and a six-speed transmission. This later model is **not** covered in this Manual.

2 O-ring chain: general

1 In an attempt to prolong chain life, many machines are fitted with chains in which lubricant is sealed into the bearings by fitting small O-rings between the inner and outer side plates of each link. The CB250 RSD-C model is fitted with such a chain; its size is 520 ($\frac{5}{8}$ x $\frac{1}{4}$ in), the chain being an endless type (no connecting link) 98 links in length. Removal, refitting and adjustment are as described in Routine Maintenance and Chapter 5.

2 While the standard chain is endless, O-ring chains of good quality are now available with connecting links from aftermarket suppliers; thus greatly easing the task of removal and refitting for cleaning purposes. If such a chain is fitted, take care that the four O-rings are positioned correctly on refitting the connecting link and its side plate, and that there is no space between the side plate and the spring clip. Be careful to fit the spring clip correctly, as described in Chapter 5.

3 O-ring chains must be lubricated to prevent wear between the sprocket teeth and the rollers and to prevent the O-rings from drying out; this should be done whenever the chain appears dry, which may mean a daily operation in wet weather. The interval given in Routine Maintenance is the absolute maximum for the purposes of chain lubrication. Use only SAE 80 or 90 gear oil; this can be applied easily via the flexible spout of the squeeze pack in which it is usually supplied. Do not forget to apply some oil to the O-rings. **Warning**: aerosol lubricants may be used **only** if clearly marked as being suitable for use with O-ring chains; the propellant used in many aerosols will attack the O-rings, thus allowing the lubricant to escape and premature chain wear to occur.

4 The chain should be renewed when it has reached the limit of its adjustment, as indicated by the red portion of the wear indicator label attached to each adjuster. If these can no longer be seen, attempt to pull the chain backwards off the rear sprocket; if, when the chain is correctly adjusted, any link can be pulled far enough off the sprocket to expose the whole of the sprocket tooth, the chain is worn out and must be renewed.

5 The chain can be cleaned while in place on the machine. Use only a soft brush and paraffin (kerosene) to remove all dirt from the rollers and O-rings. Do not use a pressure washer, steam cleaner or any strong solvent which will damage the O-rings.

3 Fuel filter: general

1 In addition to the filter described in Chapter 2, the fuel tap is fitted with a separate filter bowl which can be unscrewed to expose a second gauze filter in the base of the tap body. This must be cleaned every 4 months or 3600 miles (6000 km).

2 Switch the tap to the 'Off' position and unscrew the filter bowl using a close-fitting ring spanner. Carefully prise out the sealing O-ring and withdraw the filter gauze, which should be cleaned using only a soft brush. If traces of water or of excessive dirt are found in the fiilter bowl or gauze the tank and tap should be removed and flushed out as described in Chapter 2.

3 On refitting, ensure that the filter gauze is correctly located and press the O-ring into place to retain it. Do not overtighten the filter bowl; note that a torque setting of 0.3 – 0.5 kgf m (2 – 3.5 lbf ft) is specified. If the filter bowl is leaking, renew the sealing O-ring; overtightening the bowl in an attempt to cure a leak will only worsen the problem.

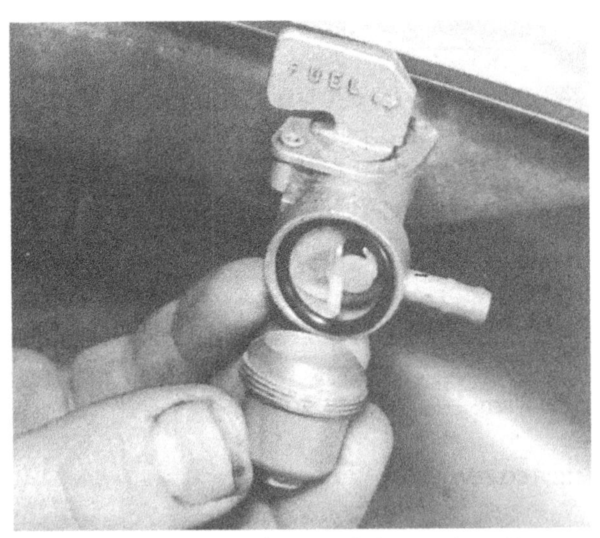

3.2 Unscrew filter bowl and remove O-ring to release filter gauze

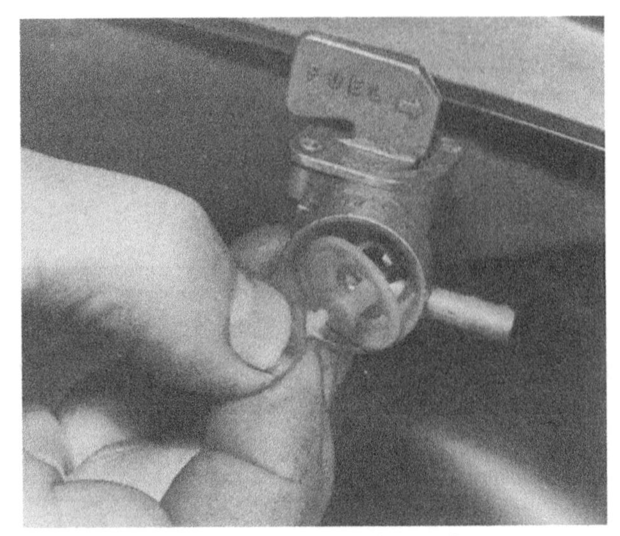

3.3 On refitting ensure gauze is located correctly on tap projection

4 Engine modifications: general

1 With reference to the various operations described in Routine Maintenance and Chapter 1, the tasks of removing and refitting the cylinder head cover and the right-hand outer casing are simplified by the absence of all components relating to the kickstart and automatic decompressor mechanisms. Note that the apertures left in the cover and casing are sealed by blanking plugs; in all other respects the engine/gearbox unit is the same. There is, of course, no decompressor cable to be adjusted on this later model.
2 On removing the left-hand outer casing, note that it can be removed either with the starter motor and drive or independently of it. In the former case, which is much quicker and easier in the long run, refer to Section 5 of this Chapter; in the latter case proceed as follows, referring to Section 5 if necessary.
3 Remove the outer cover complete with the solenoid and withdraw the thrust washer, starter clutch, coil spring and the starter drive shaft and gear. Unscrew both Allen screws from inside the aperture thus exposed and remove the three hexagon-headed screws securing the starter reduction gear housing to the left-hand outer casing. All additional work having been completed, it is now possible to remove the casing as described in Section 11 of Chapter 1. Be very careful not to damage the casing as it is withdrawn; in addition to the powerful drag of the alternator magnets the casing may stick on the two locating dowel pins or on the starter reduction gear shaft. **Do not** attempt to lever away the casing; use only gentle taps from a soft-faced mallet to release it, and be careful to press in the gear selector shaft so that it is not dislodged as the casing is removed. Note the presence of a thrust washer on the selector shaft end; this may stick to the casing.
4 On reassembly, refer to Section 39 of Chapter 1 and Section 5 of this Chapter. Apply a smear of molybdenum disulphide grease to the drive shaft bearing surfaces and helical splines before refitting.

4.1a Automatic decompressor mechanism has been removed ...

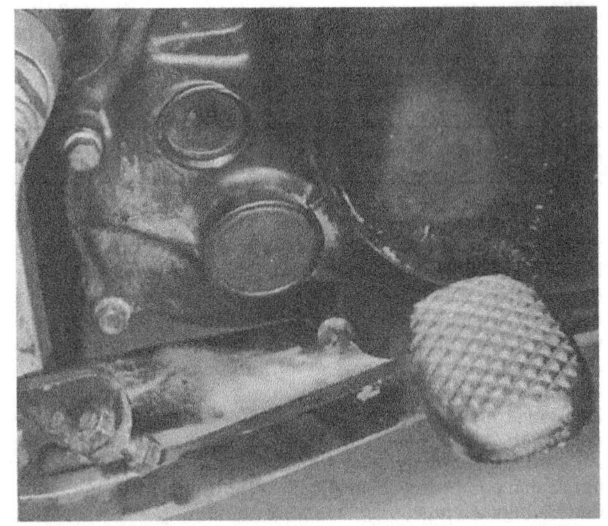

4.1b ... also kickstart mechanism – note blanking plugs sealing shaft apertures

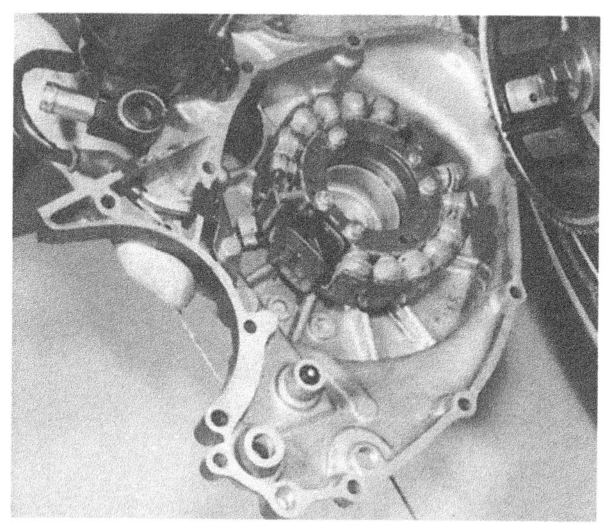
4.2 Left-hand outer casing can be removed with or without starter motor

4.3 Remove starter drive shaft components to expose two hidden Allen screws

4.4 Check selector mechanism is correctly positioned before refitting outer casing

5 Starter motor and drive mechanism – removal and refitting

1 While the solenoid and engaging mechanism can be removed separately, as can the starter clutch, drive shaft and gear, the starter motor itself cannot be removed until the left-hand outer casing has first been withdrawn. It is easiest to remove the casing and starter motor as a single unit and then to separate them, as described below.
2 Referring to Routine Maintenance and Chapter 1 where necessary, drain the engine oil, remove the gearchange linkage from the selector shaft and withdraw the gearbox sprocket cover. Remove the side panels, the seat and the fuel tank and disconnect the battery earth lead at its negative (–) terminal. Disconnect the starter motor lead at its terminal on the motor body. Tracing them from the engine unit up to the connectors joining them to the main wiring loom, disconnect the alternator and starter solenoid leads and release them from any clamps or ties securing them to the frame.
3 Using a pair of pliers, release its securing clip and pull the breather hose off its stub on the reduction gear housing. Unscrew the three mounting screws and withdraw the starter mounting bracket from the motor right-hand end.
4 Remove the three hexagon-headed screws and withdraw the outer cover complete with the solenoid. Tap the cover very gently with a soft-faced mallet to break the gasket seal and be careful to disengage the fork from the starter clutch as the cover is withdrawn; note the presence of a thrust washer on the drive shaft end. Remove the thrust washer, the starter clutch, the coil spring and the starter drive shaft and gear.
5 Unscrew the left-hand Allen screw from the back of the recess thus exposed and remove the hexagon-headed screws from around the periphery of the left-hand outer casing. Tap the casing gently with a soft-faced mallet to break the gasket seal and pull it away, being careful to press inwards on the selector shaft so that it is not dislodged. Note the presence of a thrust washer on the selector shaft end; this may stick to the casing and must be refitted on the selector shaft so that it is not lost. Due to the presence of two locating dowels and to the powerful pull of the alternator magnets the casing will be difficult to remove; be careful not to damage any component and **never** resort to levers in an attempt to remove it. Note the rubber seal between the reduction gear housing and the crankcase wall.
6 Remove the two long screws and withdraw the motor as a single unit from the reduction gear housing; note carefully the exact number of thrust washers fitted to the motor shaft and the order in which they are fitted. The motor can then be dismantled for checking, if required, as described later in this Chapter.
7 Remove the single Allen screw and the three hexagon-headed screws securing the reduction gear housing to the left-hand outer casing, then separate the two. On the machine featured in the photographs this proved to be difficult as the reduction gear shaft was a very tight fit. Support the outer casing on wooden blocks and use a soft-faced mallet to tap the housing downwards until it is released. Note which way round the double reduction gear is fitted before removing it from the shaft.
8 On reassembly, ensure that all thrust washers are refitted correctly on the starter motor shaft and that the raised tangs of the left-hand washer are engaged correctly around the boss of the reduction gear housing. Apply molybdenum disulphide grease to the seal and needle roller bearing in the housing, fit a new O-ring to the motor body and insert the motor into the housing so that the stamped line in the motor body aligns with that on the housing. Tighten the two motor retaining screws securely.

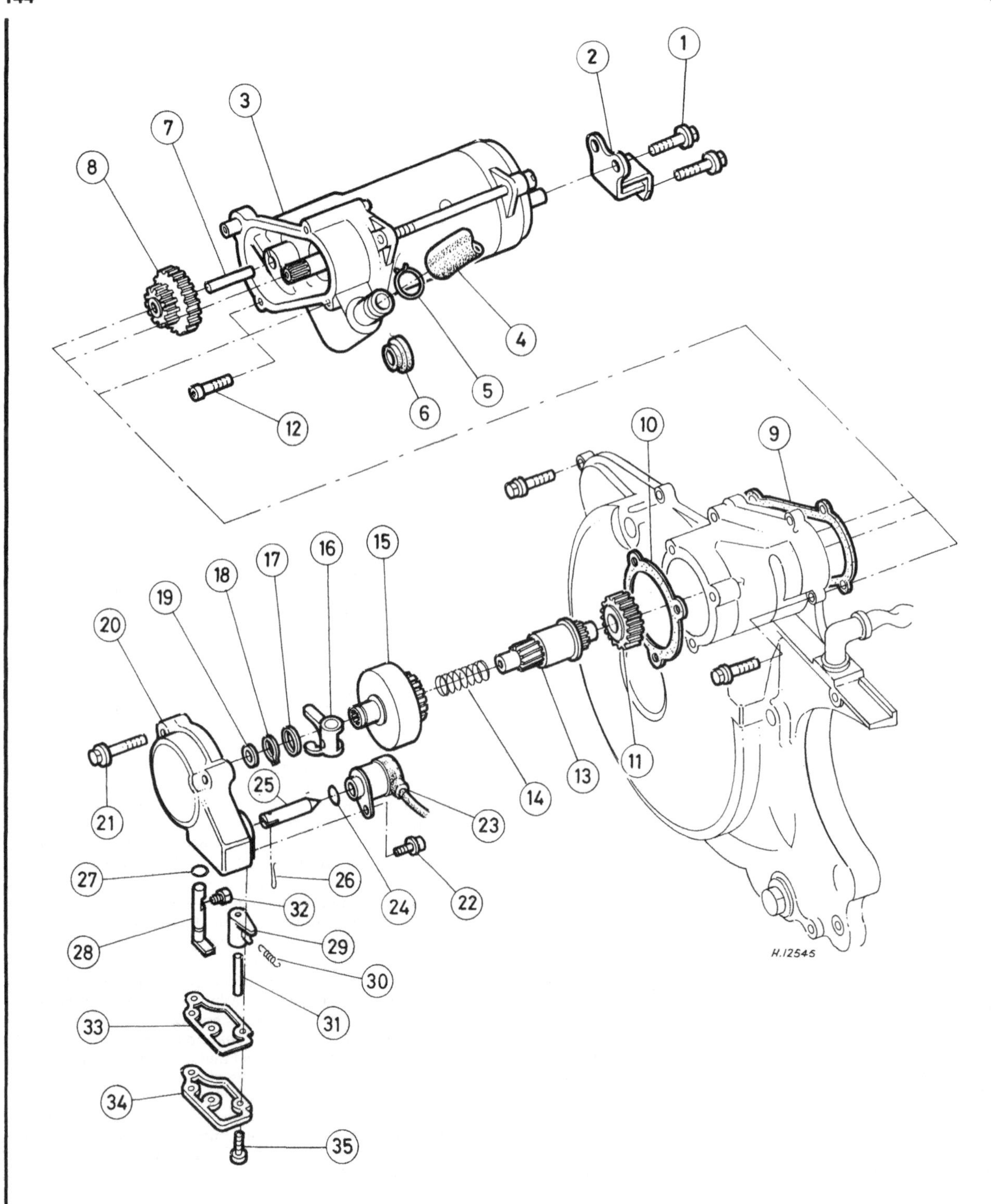

Fig. 7.1 Starter drive train

1 *Bolt – 2 off*
2 *Starter motor bracket*
3 *Starter motor*
4 *Breather hose*
5 *Hose clip*
6 *Seal*
7 *Reduction gear shaft*
8 *Reduction gear*
9 *Gasket*
10 *Gasket*
11 *Starter drive gear*
12 *Allen screw – 2 off*
13 *Starter drive shaft*
14 *Coil spring*
15 *Starter clutch*
16 *Engaging fork*
17 *Washer*
18 *Circlip*
19 *Washer*
20 *Outer cover*
21 *Bolt – 3 off*
22 *Screw*
23 *Solenoid*
24 *O-ring*
25 *Solenoid plunger*
26 *R-clip*
27 *O-ring*
28 *Pivot shaft*
29 *Starter lock cam*
30 *Lock cam spring*
31 *Pivot pin*
32 *Locking bolt*
33 *Gasket*
34 *Cover*
35 *Screw – 2 off*

9 Apply molybdenum disulphide grease to the bore of the reduction gear before refitting it to its shaft; ensure that the gear is the correct way round with the larger pinion engaging the motor shaft, and that it is free to rotate easily on its shaft. Fit the locating dowel to the housing gasket surface and use grease to stick a new gasket in place before fitting the motor assembly to the left-hand outer casing. Refit and tighten securely the four mounting screws.
10 Fit the two locating dowels to the crankcase gasket surface and use grease to stick a new gasket in place, then stick the rubber seal to its recess in the upper crankcase. Check that the selector shaft is correctly refitted, with its thrust washer, and grease the shaft splines to prevent damage to the oil seal. Offer up the outer casing/starter motor assembly and press it into place on the crankcase, ensuring that nothing is disturbed and that the cover seats securely and without stress. It may be necessary to seat the cover on the dowels with one or two gentle taps from a soft-faced mallet. Refit and tighten securely, in a diagonal sequence from the centre outwards, the retaining screws.
11 Apply a thin smear of molybdenum disulphide grease to both bearing surfaces and to the helical splines of the drive shaft then place the drive gear against the crankcase wall while the shaft is refitted. Refit the coil spring, the starter clutch and the thrust washer on the drive shaft and check that the clutch moves easily on the shaft helical splines to engage with the teeth on the alternator rotor.
12 Use grease to stick a new gasket in place on the outer casing, pull outwards as far as possible the engaging fork and refit the outer cover. Ensure that the fork ends are engaged between the starter clutch machined shoulder and the thrust washer. Tighten the three retaining screws securely.
13 Refit the breather hose to the stub on the reduction gear housing and secure the hose with its clip. Connect the starter motor lead to the terminal on the motor body, smear silicone grease over the terminal and refit the rubber cap. Refit the mounting bracket to the starter motor right-hand end and tighten the three mounting screws securely.
14 Connect the alternator and starter solenoid leads to the main loom and use the clamps or ties provided to secure the leads to the frame out of harm's way, then connect the battery earth lead to it's negative (–) terminal. Refit the fuel tank, the seat and the side panels, refill the engine with oil and refit the gearbox sprocket cover and gearchange linkage.

5.2 Disconnect starter motor lead at terminal on motor body

5.3a Release securing clip as shown and pull breather hose off its stub

5.3b Starter motor right-hand mounting is held by three screws

5.4a Remove three screws to release outer cover and solenoid ...

5.4b ... do not lose thrust washer on drive shaft end

5.4c Remove starter clutch, followed by ...

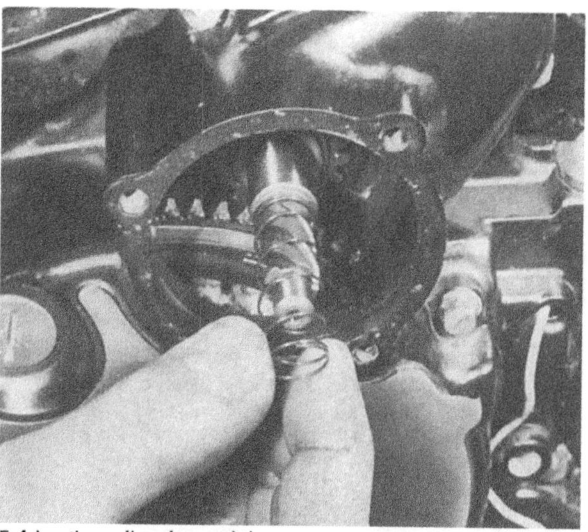

5.4d ... the coil spring and the ...

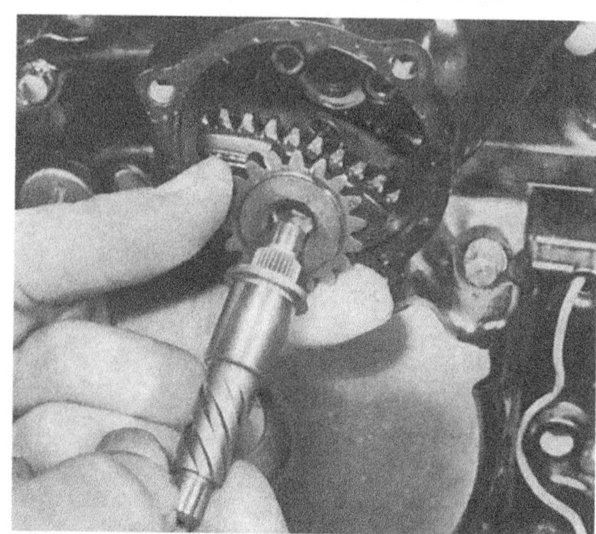

5.4e ... drive shaft and gear

5.5 Remove Allen screw shown to release outer casing from crankcase

5.6 Note carefully exact number and position of shims on starter motor shaft

5.7 Remove reduction gear housing from outer casing to release reduction gear

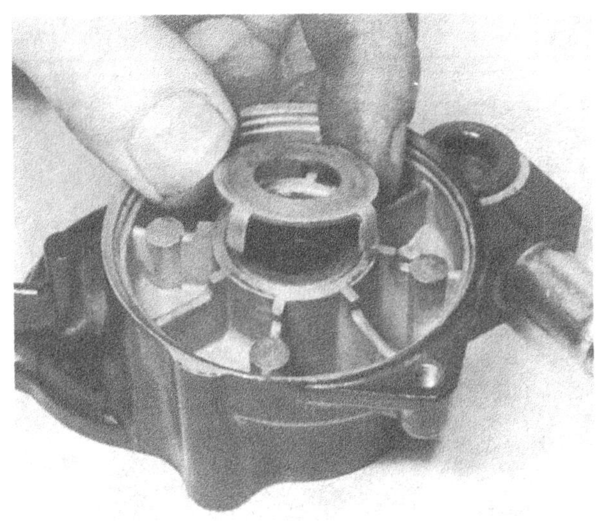
5.8a Raised tangs of left-hand thrust washer must fit as shown

5.8b Refit O-ring to motor body – do not omit shims

5.8c Align stamped line in motor body with raised mark on gear housing as shown

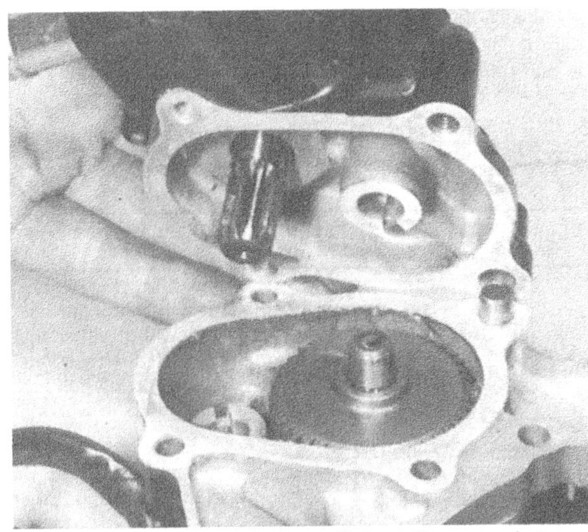
5.9 Grease reduction gear before refitting – fit new gasket on locating dowel

5.10 Do not omit rubber seal around crankcase breather passage

5.11a Grease starter drive shaft bearing surfaces and splines before refitting

5.11b Do not omit thrust washer from drive shaft end

5.12 Pull engaging fork out as far as possible and fit as shown to starter clutch

6 Starter drive and engaging mechanism: dismantling, examination and reassembly

1 All the components of the starter drive train are dismantled during the course of removing the starter motor itself, as described in the previous Section. Apart from removing the circlip from the starter clutch so that the thrust washer can be renewed if worn, nothing more can be dismantled; the clutch itself is crimped to form a sealed unit.
2 Carefully check all components for obvious signs of wear, reassembling them and feeling for free play if in doubt. Any component that is worn or damaged should be renewed immediately; pay particular attention to highly stressed areas such as the drive shaft helical splines. Wear is inevitable on the teeth of the starter clutch and alternator rotor, but can be largely ignored unless it reaches the point where the engagement of the starter is affected.
3 Check that the starter clutch can move easily on the drive shaft helical splines, renewing one or both components if movement is rough and if any tight spots are found. The clutch itself is tested by holding the centre and rotating the outer body; it should rotate smoothly anti-clockwise (viewed from the left-hand side) but should lock immediately when rotated clockwise. If this is not the case, the clutch must be renewed.
4 To dismantle the engaging mechanism, remove the two screws retaining the small rectangular cover, then remove the cover and its gasket. Remove the bolt securing the engaging fork to its pivot shaft and slide out the pivot shaft noting the O-ring fitted to it, then remove the fork. Lift out the locking cam pivot pin, disengage the cam return spring from its post and lift out the cam. Withdraw the R-clip from the solenoid plunger, unscrew the solenoid retaining screw and remove the solenoid, noting its O-ring, followed by the plunger itself.
5 The engaging mechanism is tested, with the solenoid, as described later in this Chapter. Check all components for wear, particularly the engaging fork claw ends and the mating surfaces of the fork pivot shaft and locking cam. Renew any worn or damaged items.
6 On reassembly, fit new O-rings to the solenoid and to the fork pivot shaft. Insert the plunger tapered end into the solenoid and refit the two to the outer cover, tightening the screw securely. Refit the R-clip as shown in the accompanying photograph.
7 Hook its return spring on to its post and position the locking cam against the plunger R-clip, then insert the pivot pin to retain it. Oil the fork pivot shaft, position the engaging fork in the outer cover and press in the pivot shaft to retain it. Apply thread locking compound to its threads, then refit and tighten securely the fork/shaft locking bolt. Check that all components move easily then refit the small cover and its gasket, tightening the two screws securely.

7 Ignition system: modifications

1 A different pulser assembly is fitted which no longer has identifying letters to be matched on the renewal of either the rotor (ATU) or the pulser coil. In all other respects the two units are as described in Chapter 3. Similarly the ignition HT coil has been replaced by a different item but no alterations in working procedure are necessary.
2 It should be noted that the internal resistances of both the ignition HT coil and pulser coil may be slightly different. When testing either component, the results obtained should be similar to those given in the relevant Sections of Chapter 3, but should not be regarded as conclusive proof of a unit's condition until confirmed by a competent Honda Service Agent using the correct service tools.

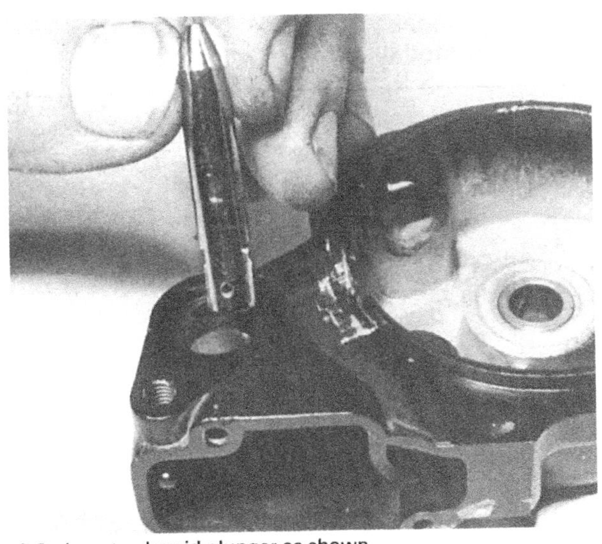

6.6a Insert solenoid plunger as shown ...

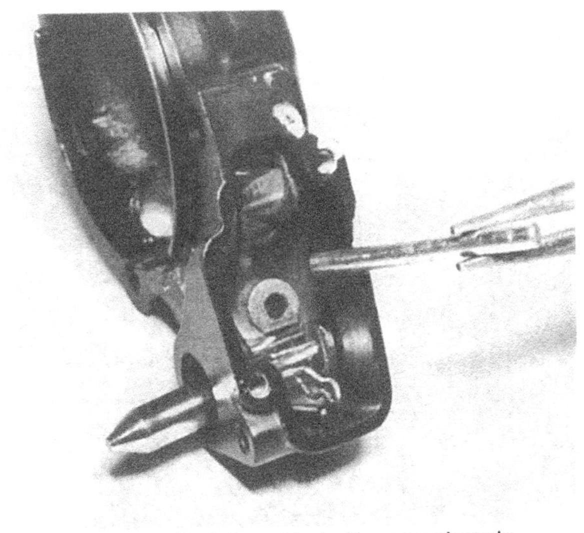

6.6b ... and refit R-clip, followed by locking cam pivot pin

6.6c Locking cam is refitted as shown ...

6.6d ... to engage with R-clip

6.7a Oil pivot shaft and new O-ring before refitting

6.7b Apply thread locking compound to threads of fork locking bolt

6.7c Check for correct operation before refitting small cover

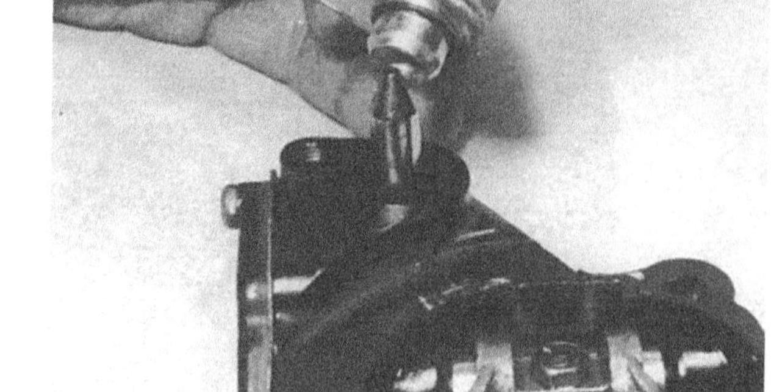

6.7d Always renew O-ring when refitting solenoid

8 Front forks: dismantling, examination and reassembly

1 The forks fitted to the later model differ in having two-piece springs, drain plugs, and separate bushes.

2 To dismantle the forks, remove them from the machine as described in Chapter 4, then remove the top bolt as described but note carefully which way up each spring is fitted to ensure correct reassembly. Note also the thick washer between the two springs. Drain the fork oil.

3 Remove the damper rod Allen bolt as described in Chapter 4, also the dust seal and the oil seal retaining circlip, then clamp the fork lower leg by its spindle lug in a vice with padded jaws,

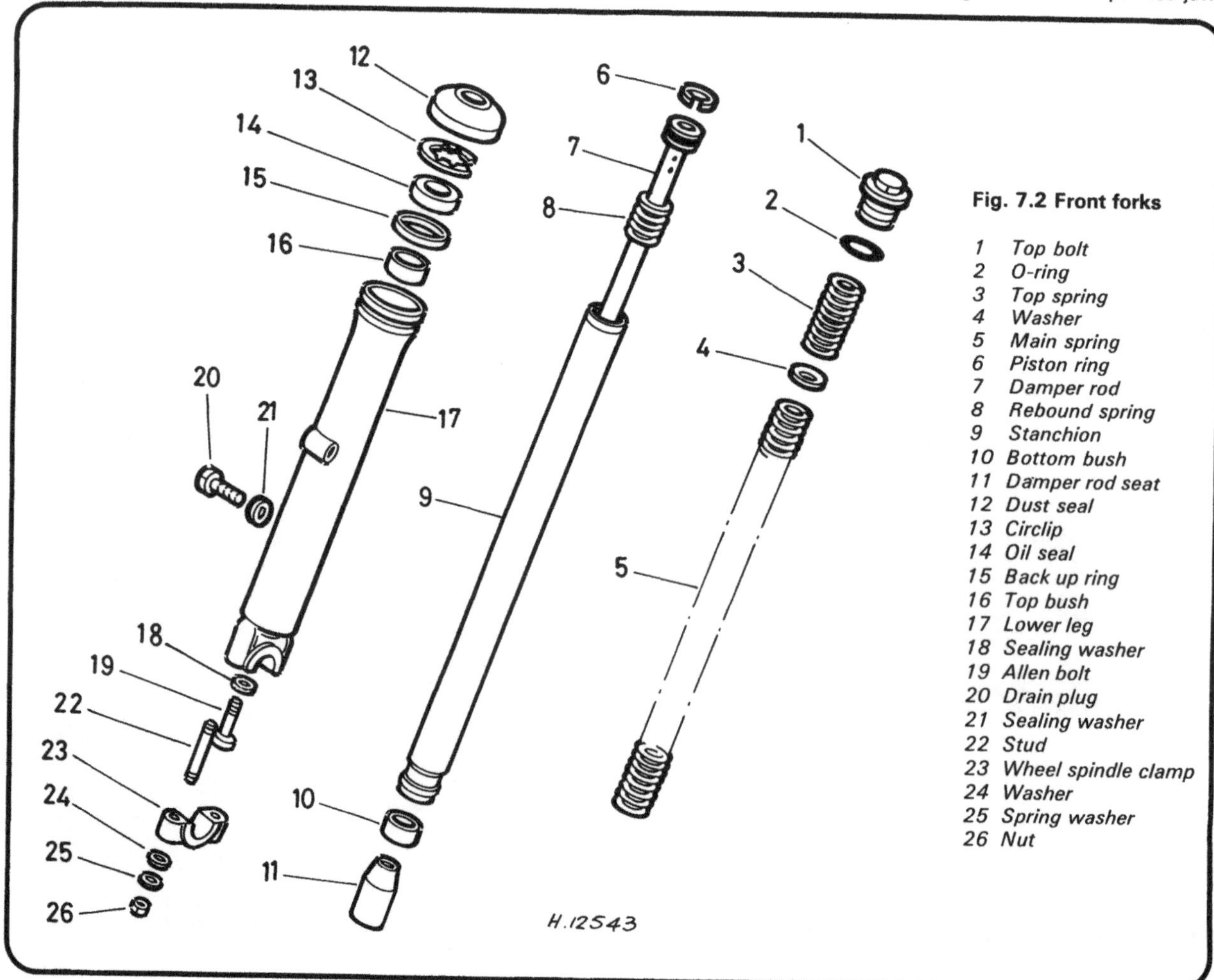

Fig. 7.2 Front forks

1 Top bolt
2 O-ring
3 Top spring
4 Washer
5 Main spring
6 Piston ring
7 Damper rod
8 Rebound spring
9 Stanchion
10 Bottom bush
11 Damper rod seat
12 Dust seal
13 Circlip
14 Oil seal
15 Back up ring
16 Top bush
17 Lower leg
18 Sealing washer
19 Allen bolt
20 Drain plug
21 Sealing washer
22 Stud
23 Wheel spindle clamp
24 Washer
25 Spring washer
26 Nut

push the stanchion fully into the lower leg and pull it sharply out as far as possible. Repeat this several times, using the slide hammer action of the bottom bush against the top bush to drive out the oil seal.

4 With the stanchion removed the oil seal, back-up ring and the top bush can be slid off the stanchion upper end. The bottom bush is split so that it can be sprung apart and eased out of its recess and off the stanchion lower end.

5 Check all fork components for wear as described in Chapter 4 with the following exceptions. Since measurements are not provided with which the fork spring condition can be assessed, these can be checked only by comparison with new components. To check the bushes for wear, examine the bearing surface of each one; if the teflon is worn away so that the copper material appears over more than ¾ of the whole bearing surface, that bush must be considered worn out. It would be best to renew all the bushes together to preserve equal fork performance. Make a final check by refitting the bushes to the stanchion and inserting the assembly into the lower leg; little or no free play should be discernible.

6 On reassembly insert the damper rod, complete with piston ring and rebound spring, into the stanchion and press it down so that it projects fully from the stanchion lower end, then refit the damper rod seat. Check that the bottom bush is correctly seated, smear fork oil over the stanchion and bush then insert the assembly into the lower leg. Press the stanchion fully down to centralise the damper rod seat and refit the Allen bolt as described in Chapter 4.

7 Smearing it with fork oil, slide the top bush down over the stanchion followed by the seal back-up ring and press the bush into its recess in the fork lower leg. Use a hammer and suitable drift to tap the seal seat and bush fully into place. Smear the stanchion with fork oil to protect the seal lips and slide down the oil seal with its marked surface upwards.

8 Press the seal squarely into the lower leg as far as possible by hand only, then use as a drift a length of tubing of the same inside and outside diameters as the seal hard outer edge to tap the seal into place until the circlip groove is exposed. Refit the circlip and dust seal, fill the fork leg with the specified amount of oil, then refit the fork springs, with their thick washer, and the fork top bolt.

9 Front brake master cylinder: modifications

Although a modified type of master cylinder is fitted, with slight alterations to the layout of the piston assembly seals, there is no difference in working procedure. Note carefully which way round the seals are fitted on removal, and ensure that the new ones are fitted in the same way. Similarly, while a new type of stop lamp front switch is fitted, it is retained by a single screw and is fitted as shown in Chapter 5.

10 Rear wheel bearings and cush drive: modifications

1 The rear wheel hub is different in layout to the earlier model, as shown in the accompanying photographs. A retainer is threaded into a large boss on the hub left-hand side and must be unscrewed using a tool that can be fabricated as shown in the accompanying illustration. Once the retainer is removed, the oil seal behind it can be levered out of its housing or driven out with the bearing. Since the central spacer is no longer flanged at one end there is no particular order in which the bearings are to be removed.

2 On reassembly, ensure that the retainer is tightened fully and securely into the hub, then prevent it from unscrewing by staking it with a hammer and punch at each of the four holes.

3 The modified cush drive assembly is shown in the photographs. It consists of six double rubber blocks arranged in the hub so that vanes cast in the rear face of the sprocket carrier can fit into them. To dismantle the cush drive, peel off the rubber cover and remove the large circlip and the thrust washer beneath. It should then be possible to lift away the sprocket assembly but if it is a tight fit, tap the sprocket away by passing a wooden drift through the wheel spokes to bear on the sprocket from the rear. The sprocket can be unbolted from its carrier without disturbing the cush drive, if required.

4 Check that the rubber blocks are not cracked, perished or excessively worn. Renew all six as a set, even if only one is worn or damaged. Clean the hub boss and apply a smear of grease to prevent corrosion, and check the condition of the O-ring around its base, renewing it if worn or damaged.

5 On reassembly, position the rubber blocks as shown in the accompanying photographs and apply a suitable lubricant to each one to aid reassembly. The sprocket carrier will be a tight fit, especially if new rubbers have been fitted. Refit the thrust washer, circlip and the rubber cover.

10.3 Sprocket is bolted to a separate carrier, which engages with cush drive blocks in hub

10.5 Fit new cush drive blocks as shown

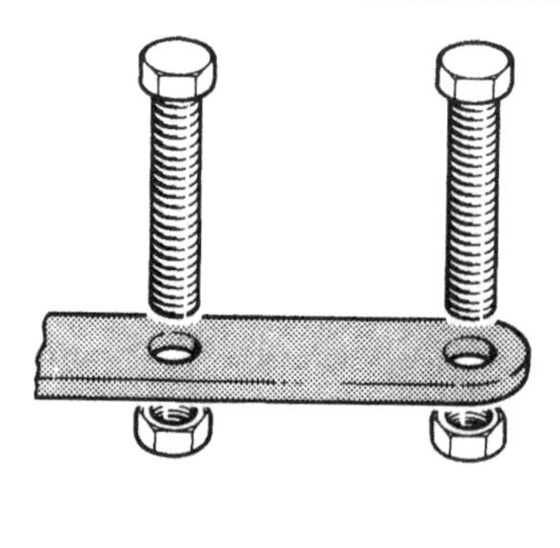

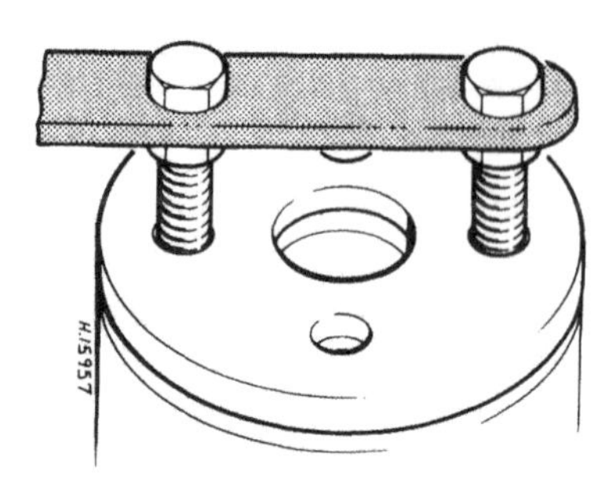

Fig. 7.3 Fabricated peg spanner for removing and refitting wheel bearing retainer

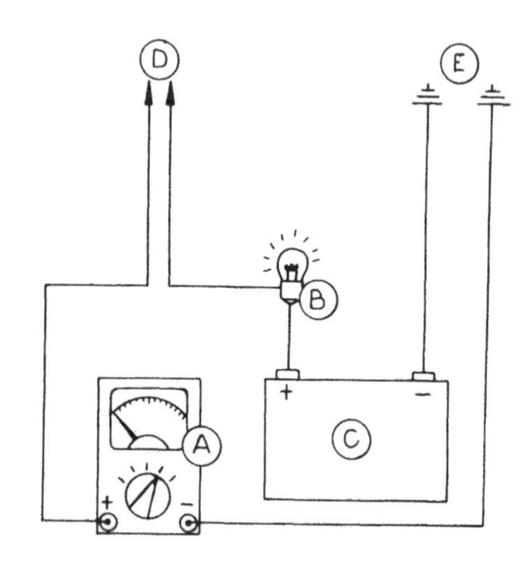

Fig. 7.4 Continuity test circuits

A Multimeter
B Bulb
C Battery
D Positive probe
E Negative probe

11 Starter motor circuit: general

1 As a starter motor of the pre-engaged type is employed, the circuit comprises not only the starter motor, switch and relay, but also a starter solenoid with a separate relay to energise it. The neutral indicator lamp circuit is connected to the starter button feed wire to ensure that the starter will not work unless the gearbox is in the neutral position.

2 In addition to this safety lock-out, a switch is fitted in the clutch handlebar lever to ensure that the starter will not work unless the lever is pulled in. A diode is fitted between the two lock-out circuits to allow the clutch to override the neutral lock-out. This means that the machine can be started in gear provided that the clutch lever is first pulled in.

3 In the event of a starter fault, isolate the problem in a logical manner. If the starter motor is turning over, but not turning the engine, the fault must lie in the engaging or drive mechanism. Check the solenoid action as described later in this Chapter to eliminate the fault.

4 If the starter does not rotate, or only very slowly, check first the battery, then the ignition switch. For the purposes of a quick check, these can be eliminated by checking the brightness of the lights and correct operation of other electrical components. If the battery is fully charged and the ignition switch in working order, check the starter button, followed by the neutral and clutch switches and the diode between them.

5 If the fault is still not found, check the starter relay and the wiring and cables, followed by the starter motor itself. Refer if necessary to the relevant Sections of this Chapter and of Chapter 6 for details of testing the individual components.

6 In the event of a major and obvious fault such as the motor not rotating it will be sufficient to use a multimeter or a dry battery and bulb test circuit, as shown in the accompanying illustration, to isolate the fault. In the event of an intermittent fault, or if the starter motor only rotates slowly with a fully charged battery, the fault will be more difficult to trace and will require careful checking with a multimeter. A switch, for example, may be so corroded or fouled with dirt that it is creating a considerable resistance, thus robbing the motor of most of the power available.

7 Faults in the starter drive train can be found only by dismantling and checking the components, as described in Sections 5 and 6 of this Chapter.

12 Starter motor: overhaul and testing

1 The starter motor can be removed only after the left-hand outer casing has been removed, or with it, as described in Section 5 of this Chapter.

2 Once the two screws have been removed to release the motor from the reduction gear housing, the end cover and O-ring can be lifted away. Note that shims are fitted to the motor left-hand end only; note the exact number and positon of these before disturbing them.

3 Carefully ease the springs out of the brush holders to release the brushes then lift off the brush holder plate. Disengage the field coil brush leads from the plate to release it. Push the armature out of the motor body. Unscrew the terminal retaining nut, withdraw the four washers and the O-ring and remove the terminal bolt. This will release the field coil brush assembly. The plastic insulator and nylon guide can be removed if necessary.

4 Measure the length of each brush; they are worn out if reduced to 6.5 mm (0.26 in) or less. Check that the brushes are not chipped or otherwise damaged and that they make good contact with the commutator. Note that since the two field coil brushes are attached to a single clip and the leads to the remaining two brushes are crimped on to the holder plate, the brushes are only available as a set, with the clip and holder plate.

5 If a spring balance or similar is available check that the springs exert the correct pressure on the brushes, and that each

brush is free to slide easily in its holder. The springs are not available separately and will be renewed with the holder plate.

6 Clean the commutator segments with a rag soaked in methylated spirits and inspect each one for scoring or discolouration. If any pair of segments is discoloured, a shorted armature winding is indicated. The manufacturer supplies no information regarding skimming and re-cutting the armature in the event of serious scoring or burning, and so by implication suggests that a new armature, ie a new starter motor assembly, is the only solution. It is suggested, however, that the advice of a vehicle electrical specialist is sought first; professional help may work out a lot cheaper.

7 Honda advise against cleaning the commutator segments with abrasive paper, presumably because of the risk of abrasive particles becoming embedded in the soft segments. It is suggested, therefore, that an ink eraser be used to burnish the segments and remove any surface oxide deposits before installing the brushes.

8 Using a multimeter set on the resistance scale, check the continuity between pairs of segments, noting that anything other than a very low resistance indicates a partially or completely open circuit. Next check the armature insulation by checking for continuity between the armature core and each segment. Anything other than infinite resistance indicates an internal failure.

9 Check the field coil brushes by testing for continuity between each brush and the terminal; no resistance should be encountered. Check also that the terminal is completely insulated from the motor body. Finally, check that continuity (ie little or no resistance) exists between the field coils, then that each coil is completely insulated from the motor body. If continuity is found between the field coils and the body there is a breakdown in insulation which means that the complete assembly must be renewed.

10 If oil is found in the starter motor, the seal pressed into the reduction gear housing is faulty and must be renewed. Check the bearing at each end of the armature by reassembling the motor and feeling for free play. Spin each bearing and check for signs of roughness, wear, or other damage. The bearings must be renewed if at all worn, but note that neither these nor the seal are available separately; the apparent solution is to purchase a new starter motor assembly. To avoid unnecessary expense, an automotive parts supplier or specialist bearing supplier may be able to find suitable replacements. Ensure that all relevant dimensions and seal or bearing markings are noted so that the correct items can be selected; if necessary take the motor assembly to provide a pattern.

11 The right-hand bearing can be removed using a knife-edged bearing puller and is refitted using a hammer and a tubular drift such as a socket spanner which bears only on the inner race. Ensure that the armature is fully supported. The oil seal can be levered out of the reduction gear housing and the needle roller bearing drifted out once the housing has been heated to approximately 100°C in an oven. Both are refitted as described above, using a socket which bears only on the bearing outer race or seal outer edge.

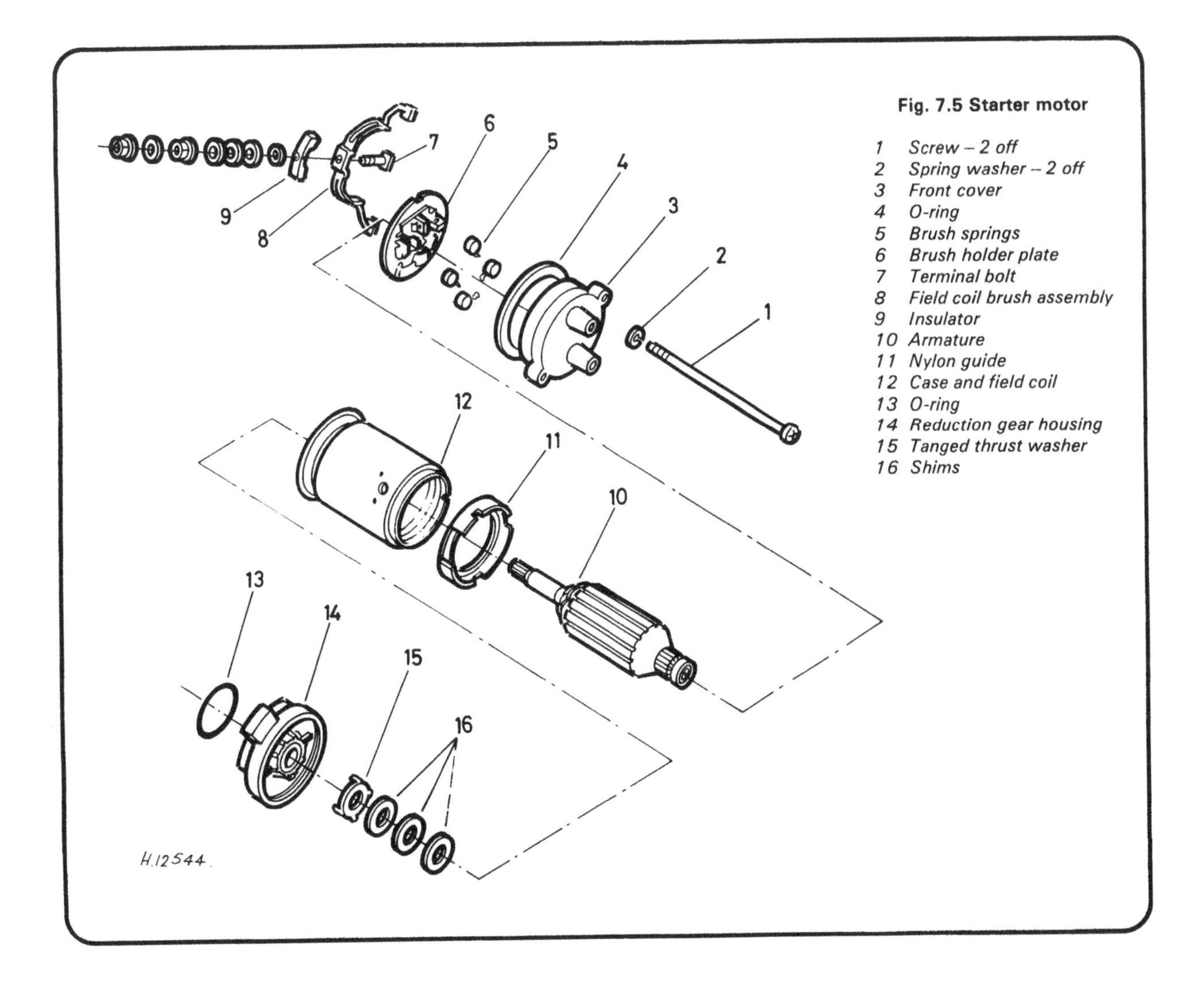

Fig. 7.5 Starter motor

1 Screw – 2 off
2 Spring washer – 2 off
3 Front cover
4 O-ring
5 Brush springs
6 Brush holder plate
7 Terminal bolt
8 Field coil brush assembly
9 Insulator
10 Armature
11 Nylon guide
12 Case and field coil
13 O-ring
14 Reduction gear housing
15 Tanged thrust washer
16 Shims

12 On reassembly, fit the nylon guide to the motor body, followed by the insulator, the field coil brush clip and the terminal bolt. Check that all components are seated correctly before fitting the sealing O-ring and washers to the terminal bolt and tightening securely its retaining nut. Insert the armature into the body.
13 Check that the brush springs are clear of the holder and the brushes withdrawn. Offer up the holder plate, press the field coil brush wires into their slots and press the plate into position, aligning it with the locating tangs in the motor body. Insert all four brushes and refit the springs. Check that each brush can slide easily in its holder before spring pressure is applied.
14 Fit a new O-ring to the motor body right-hand end and refit the end cover, aligning the stamped line on the body with the raised mark on the end cover. Do not forget to fit a new O-ring to the body left-hand end, or to refit the shims in their original positions on the armature shaft before refitting the motor to the machine as described in Section 5.
15 Note that the motor right-hand mounting bracket consists of two metal brackets joined by a bonded rubber buffer. Renew the bracket if the rubber is perished or damaged.

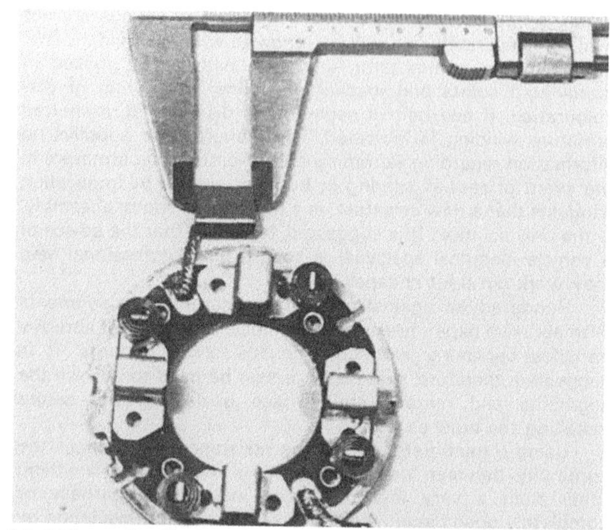
12.4 Measuring the length of a starter motor brush

12.12a Ensure nylon guide fits correctly in motor body

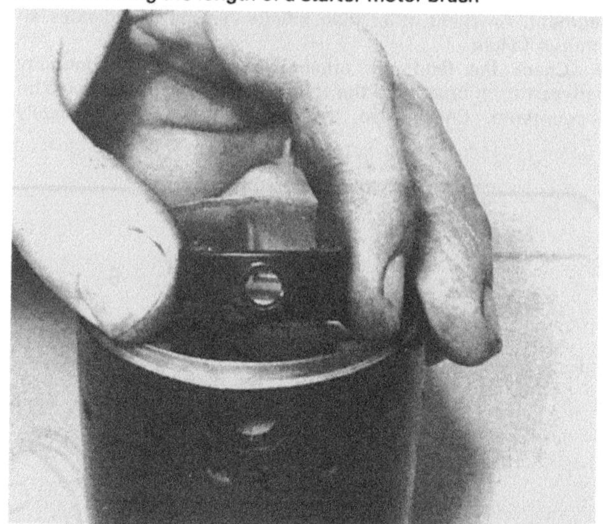
12.12b Raised shoulder on insulator locates in casing terminal bolt aperture

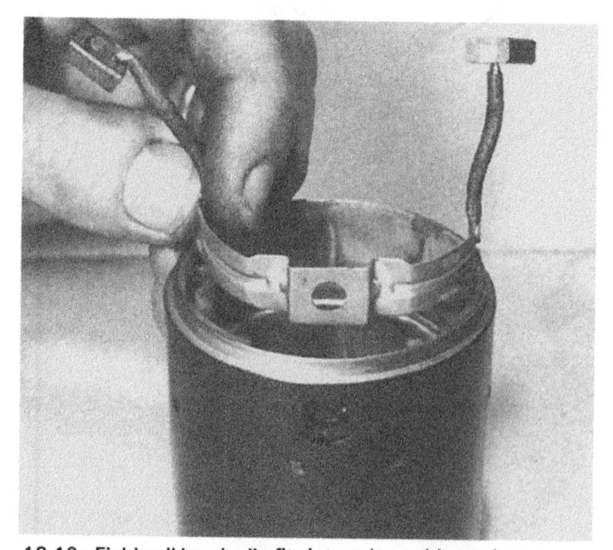
12.12c Field coil brush clip fits into nylon guide as shown

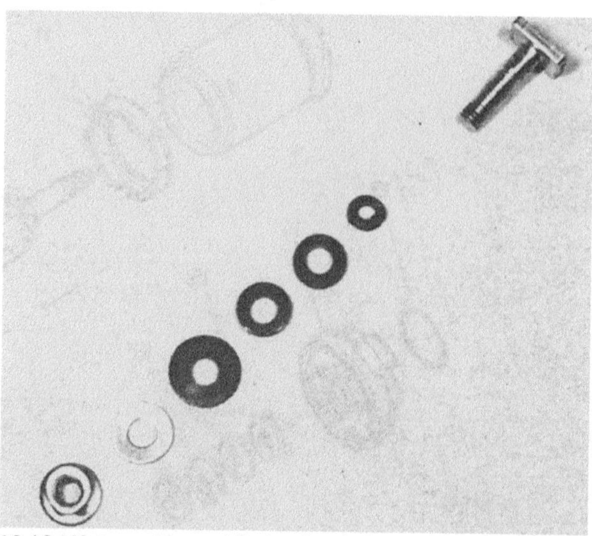
12.12d Note positions of O-ring, insulating and metal washers and retaining nut

12.12e Take care when inserting armature into motor body – magnets are very strong

12.13 Check brushes slide easily in holders before refitting springs

12.14a Renew sealing O-ring if damaged or worn

12.14b Align stamped line in motor body with raised mark on end cover

13 Starter button and relay: location and testing

1 The starter button is part of the right-hand handlebar switch cluster. To test it remove the headlamp rim and reflector assembly from the shell then trace and disconnect the connector joining the switch cluster wires to the main loom. With a multimeter set to its resistance function, check for continuity between the black and the yellow/red wire terminals when the button is pressed; if the switch is in good condition no resistance should be measurable.

2 If resistance is encountered, the switch must be renewed. Since this means the renewal of the complete switch cluster there is nothing to be lost by attempting to dismantle the switch to clean the contacts and to rectify the fault. Check that the problem is not due to a broken wire or to a wire that has been trapped to create a short circuit; this can be repaired relatively easily.

3 The starter relay is rubber-mounted beneath the seat, just behind the fuse box. Remove the seat, the right-hand side panel and the fuse box to gain access to the unit, which can be identified by the two heavy duty cables leading to two of its four terminals.

4 To make a quick test of the unit, switch on the ignition and press the starter button; a distinct click should be heard as the internal solenoid closes the starter lead contact. While a silent relay may be assumed to be faulty, check first that the battery is fully charged. A partially-discharged battery may have sufficient power for all other electrical components to operate normally, but not enough to supply the very heavy current required to start the engine, although this is usually shown by a rapid series of clicks as the relay tries to operate correctly.

5 Disconnect the heavy duty starter lead at the motor terminal and connect a 12 volt test bulb between the lead and a good earth point. If the bulb lights when the starter button is pressed the motor is being supplied with power. The fault therefore lies in the motor, which must be removed for testing

and overhaul as described in Section 12 of this Chapter, and not in the relay. If the bulb does not light, the relay could be faulty and should be tested as follows.

6 Unscrewing their retaining nuts, disconnect from the relay the two heavy duty cables and the lead formed by two red wires, then disconnect the two remaining wires leading to the relay. Connect a fully charged 12 volt battery across the two wires, positive (+) terminal to the yellow/red wire and negative (–) terminal to the green/red wire, to energise the primary coil or solenoid, then use a multimeter to check for continuity across the secondary (threaded) terminals. If any resistance is measured, the relay is faulty and must be renewed.

13.3 Location of starter relay

14 Starter engaging solenoid and relay: location and testing

1 If the starter motor spins freely without engaging the engine, especially if the noise of the meshing components cannot be heard, the engaging solenoid and its relay should be checked.

2 The relay is rubber mounted on a tang projecting to the rear of the fuel tank rear mounting point. Remove the seat to gain access to the unit which is a sealed metal box with a four-pin block connector leading to it.

3 To test the unit, unplug the connector block and carefully note which wire leads to which terminal on the unit. Connect a fully charged 12 volt battery across the unit terminals of the green/red and yellow/red wires, then use a multimeter set to its resistance function to check for continuity across the terminals of the brown/red and black wires. If any resistance is encountered the unit is faulty and must be renewed.

4 To test the engaging solenoid remove as necessary the side panel and the seat until its wires can be disconnected from the main loom, then remove the small rectangular cover which is retained by two screws to the underside of the engaging mechanism outer cover. Connect a fully charged 12 volt battery across the solenoid wires, positive (+) terminal to the brown/red wire and negative (–) terminal to the green wire, to energise the solenoid so that its operation can be checked. Note that where both solenoid wires are black, care must be taken to identify correctly each wire by means of the colour coding of the wires coming from the main loom.

5 On being connected as described the solenoid should immediately pull in its plunger to engage the starter clutch on to the alternator rotor teeth. If the solenoid does not work as described it is faulty and must be renewed. Remove the outer cover and check first that the fault is genuine and not due to some other problem such as a stuck or corroded plunger.

6 To test the engaging mechanism, remove the three retaining screws and withdraw the outer cover complete with the solenoid. Energise the solenoid as described above and pull the engaging fork outwards, then try to push it in. If little resistance is encountered to normal hard pressure, dismantle the mechanism and check the locking cam and fork pivot shaft for wear, renewing any worn or damaged components.

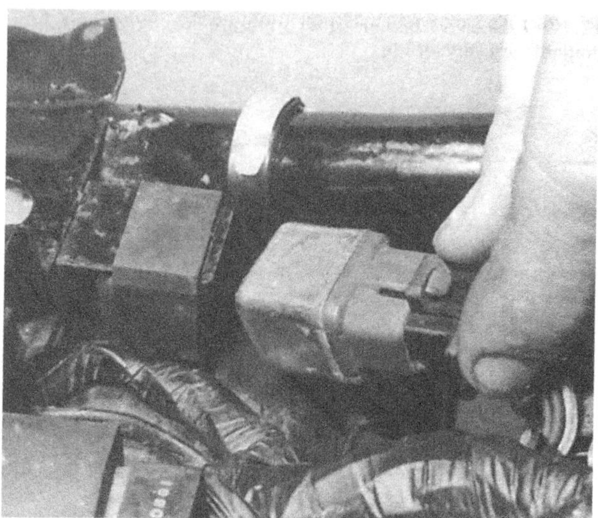

14.2 Location of engaging solenoid relay

15 Starter lockout circuits: testing

1 The neutral indicator lamp circuit is tested as described in Section 15 of Chapter 6.

2 The clutch switch is pressed into the handlebar lever clamp. To test it, use a multimeter set to the resistance function to check that continuity exists between its two terminals only when the lever is pulled in to the handlebar. If any resistance is measured, the switch is faulty and must be renewed.

3 To remove the switch, disconnect its wires and use a suitably pointed instrument to press in the locking tab so that the switch can be pressed out of the clamp after the lever has been removed. Refitting is the reverse of this, ensuring that the locking tab engages correctly to retain the switch.

4 The diode between the two circuits is a small rectangular black plastic unit with two spade terminals that is rubber mounted next to the engaging solenoid relay. Remove the seat to gain access to it.

5 To test the diode, unplug it from its connector and use a multimeter set to the resistance function to check for continuity. Connect the probes to the terminals for one reading, then reverse the probes for the second. Continuity should be found in one direction only, marked by an arrow on the unit (and shown in the wiring diagram); if continuity is found in both directions or in none, the diode is faulty and must be renewed.

16 Headlamp bulb: general

While a bulb of 12V, 60/55W rating is now employed, it is removed and refitted as described in Chapter 6. Never touch the glass envelope of a quartz halogen bulb as it is easily damaged by oil or skin acids; take great care when handling the bulb. Note also that their service life is much reduced if the supply voltage is allowed to drop.

15.4 Location of starter lockout circuit diode

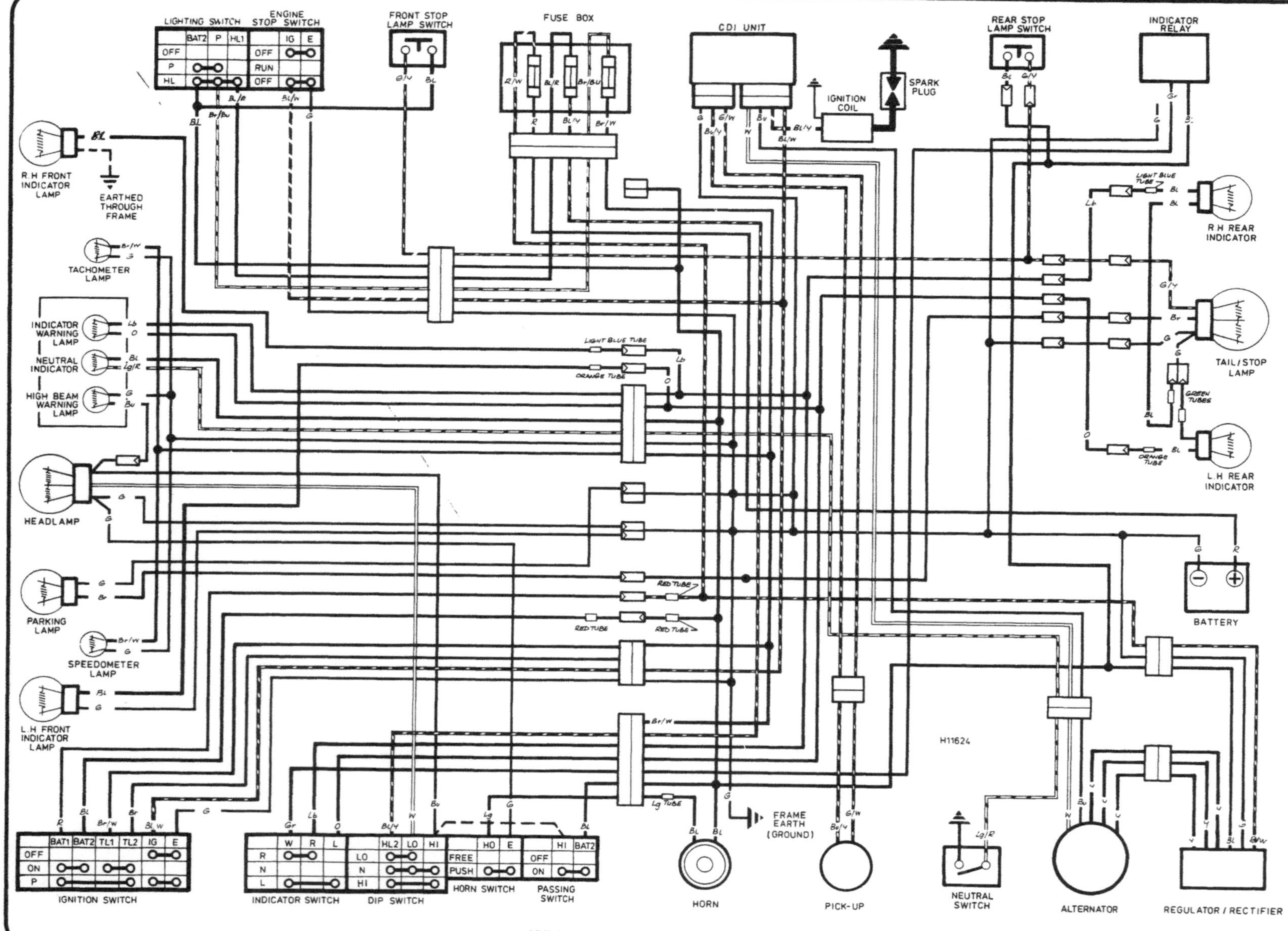

LIGHTING SWITCH
ENGINE STOP SWITCH
FRONT STOP LAMP SWITCH
FUSE BOX
CDI UNIT
IGNITION COIL
SPARK PLUG
REAR STOP LAMP SWITCH
INDICATOR RELAY
R.H FRONT INDICATOR LAMP
EARTHED THROUGH FRAME
TACHOMETER LAMP
INDICATOR WARNING LAMP
NEUTRAL INDICATOR
HIGH BEAM WARNING LAMP
HEADLAMP
PARKING LAMP
SPEEDOMETER LAMP
L.H FRONT INDICATOR LAMP
R.H REAR INDICATOR
TAIL/STOP LAMP
L.H REAR INDICATOR
BATTERY
IGNITION SWITCH
INDICATOR SWITCH
DIP SWITCH
HORN SWITCH
PASSING SWITCH
HORN
FRAME EARTH (GROUND)
PICK-UP
NEUTRAL SWITCH
ALTERNATOR
REGULATOR / RECTIFIER
H11624

Colour Code

Bl	Black
Br	Brown
Bu	Blue
G	Green
Gr	Grey
Lb	Light blue
Lg	Light green
O	Orange
R	Red
W	White
Y	Yellow

Wiring diagram – Honda CB250 RS-A

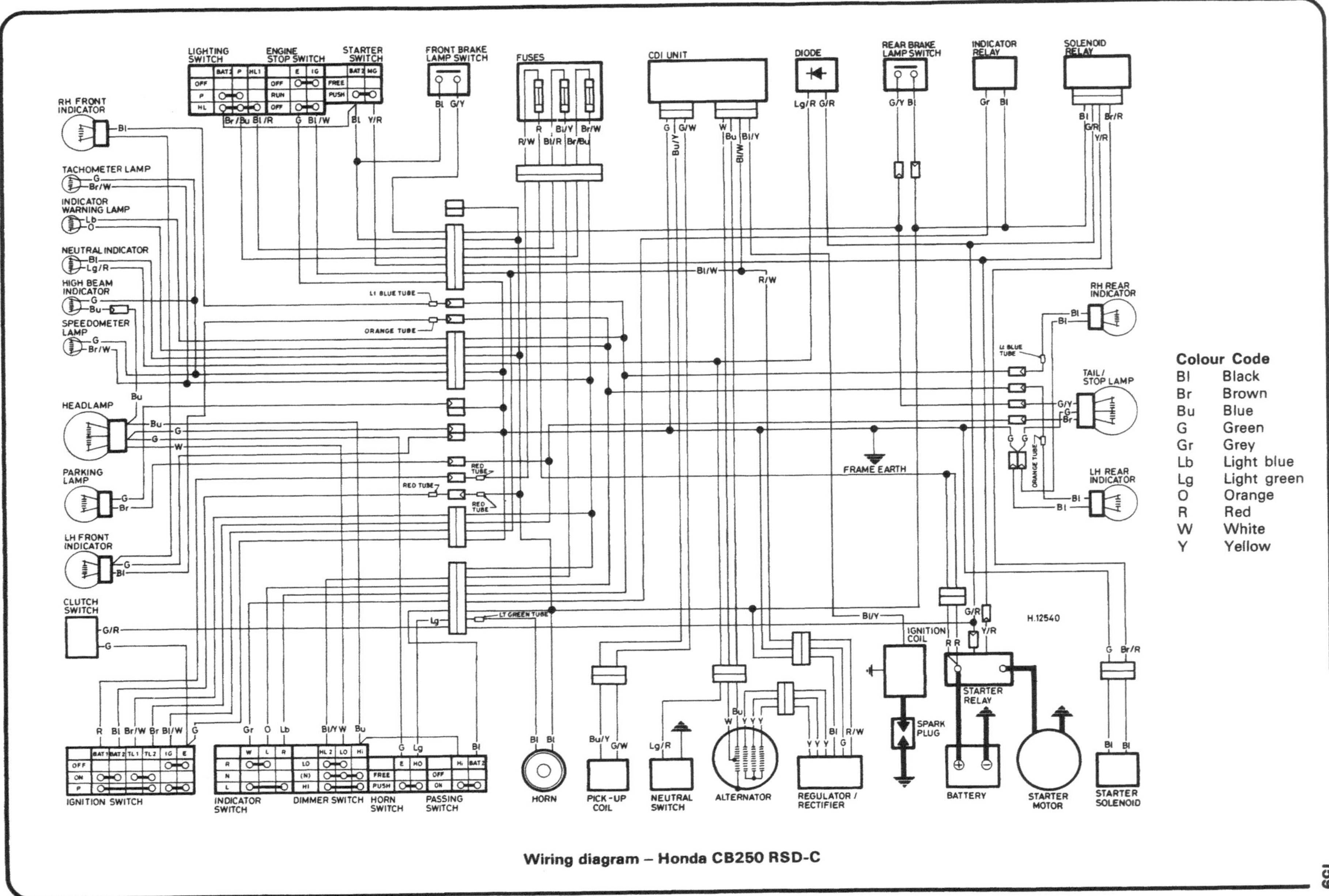

Wiring diagram – Honda CB250 RSD-C

Conversion factors

Length (distance)					
Inches (in)	X 25.4	= Millimetres (mm)	X 0.039	= Inches (in)
Feet (ft)	X 0.305	= Metres (m)	X 3.281	= Feet (ft)
Miles	X 1.609	= Kilometres (km)	X 0.621	= Miles
Volume (capacity)				
Cubic inches (cu in; in³)	X 16.387	= Cubic centimetres (cc; cm³)	X 0.061	= Cubic inches (cu in; in³)
Imperial pints (Imp pt)	X 0.568	= Litres (l)	X 1.76	= Imperial pints (Imp pt)
Imperial quarts (Imp qt)	X 1.137	= Litres (l)	X 0.88	= Imperial quarts (Imp qt)
Imperial quarts (Imp qt)	X 1.201	= US quarts (US qt)	X 0.833	= Imperial quarts (Imp qt)
US quarts (US qt)	X 0.946	= Litres (l)	X 1.057	= US quarts (US qt)
Imperial gallons (Imp gal)	X 4.546	= Litres (l)	X 0.22	= Imperial gallons (Imp gal)
Imperial gallons (Imp gal)	X 1.201	= US gallons (US gal)	X 0.833	= Imperial gallons (Imp gal)
US gallons (US gal)	X 3.785	= Litres (l)	X 0.264	= US gallons (US gal)
Mass (weight)				
Ounces (oz)	X 28.35	= Grams (g)	X 0.035	= Ounces (oz)
Pounds (lb)	X 0.454	= Kilograms (kg)	X 2.205	= Pounds (lb)
Force				
Ounces-force (ozf; oz)	X 0.278	= Newtons (N)	X 3.6	= Ounces-force (ozf; oz)
Pounds-force (lbf; lb)	X 4.448	= Newtons (N)	X 0.225	= Pounds-force (lbf; lb)
Newtons (N)	X 0.1	= Kilograms-force (kgf; kg)	X 9.81	= Newtons (N)
Pressure				
Pounds-force per square inch (psi; lbf/in²; lb/in²)	X 0.070	= Kilograms-force per square centimetre (kgf/cm²; kg/cm²)	X 14.223	= Pounds-force per square inch (psi; lbf/in²; lb/in²)
Pounds-force per square inch (psi; lbf/in²; lb/in²)	X 0.068	= Atmospheres (atm)	X 14.696	= Pounds-force per square inch (psi; lbf/in²; lb/in²)
Pounds-force per square inch (psi; lbf/in²; lb/in²)	X 0.069	= Bars	X 14.5	= Pounds-force per square inch (psi; lbf/in²; lb/in²)
Pounds-force per square inch (psi; lbf/in²; lb/in²)	X 6.895	= Kilopascals (kPa)	X 0.145	= Pounds-force per square inch (psi; lbf/in²; lb/in²)
Kilopascals (kPa)	X 0.01	= Kilograms-force per square centimetre (kgf/cm²; kg/cm²)	X 98.1	= Kilopascals (kPa)
Torque (moment of force)				
Pounds-force inches (lbf in; lb in)	X 1.152	= Kilograms-force centimetre (kgf cm; kg cm)	X 0.868	= Pounds-force inches (lbf in; lb in)
Pounds-force inches (lbf in; lb in)	X 0.113	= Newton metres (Nm)	X 8.85	= Pounds-force inches (lbf in; lb in)
Pounds-force inches (lbf in; lb in)	X 0.083	= Pounds-force feet (lbf ft; lb ft)	X 12	= Pounds-force inches (lbf in; lb in)
Pounds-force feet (lbf ft; lb ft)	X 0.138	= Kilograms-force metres (kgf m; kg m)	X 7.233	= Pounds-force feet (lbf ft; lb ft)
Pounds-force feet (lbf ft; lb ft)	X 1.356	= Newton metres (Nm)	X 0.738	= Pounds-force feet (lbf ft; lb ft)
Newton metres (Nm)	X 0.102	= Kilograms-force metres (kgf m; kg m)	X 9.804	= Newton metres (Nm)
Power				
Horsepower (hp)	X 745.7	= Watts (W)	X 0.0013	= Horsepower (hp)
Velocity (speed)				
Miles per hour (miles/hr; mph)	X 1.609	= Kilometres per hour (km/hr; kph)	X 0.621	= Miles per hour (miles/hr; mph)
Fuel consumption*				
Miles per gallon, Imperial (mpg)	X 0.354	= Kilometres per litre (km/l)	X 2.825	= Miles per gallon, Imperial (mpg)
Miles per gallon, US (mpg)	X 0.425	= Kilometres per litre (km/l)	X 2.352	= Miles per gallon, US (mpg)

Temperature

Degrees Fahrenheit (°F) $= (°C \times \frac{9}{5}) + 32$

Degrees Celsius (Degrees Centigrade; °C) $= (°F - 32) \times \frac{5}{9}$

**It is common practice to convert from miles per gallon (mpg) to litres/100 kilometres (l/100km), where mpg (Imperial) x l/100 km = 282 and mpg (US) x l/100 km = 235*

Index